REVITALIZING ELECTORAL GEOGRAPHY

Revitalizing Electoral Geography

Edited by
BARNEY WARF
University of Kansas, USA
and
JONATHAN LEIB
Old Dominion University, USA

Routledge
Taylor & Francis Group
LONDON AND NEW YORK

First published 2011 by Ashgate Publishing

Published 2016 by Routledge
2 Park Square, Milton Park, Abingdon, Oxfordshire OX14 4RN
711 Third Avenue, New York, NY 10017, USA

First issued in paperback 2016

Routledge is an imprint of the Taylor & Francis Group, an informa business

British Library Cataloguing in Publication Data
Revitalizing electoral geography.
1. Electoral geography. 2. Electoral geography--Europe.
3. Electoral geography--United States.
I. Warf, Barney, 1956- II. Leib, Jonathan.
324.6-dc22

Library of Congress Cataloging-in-Publication Data
Warf, Barney, 1956-
Revitalizing electoral geography / by Barney Warf and Jonathan Leib.
p. cm.
Includes index.
ISBN 978-1-4094-1072-0 (hbk)
1. Electoral geography. 2. Electoral geography--Case studies. I. Leib, Jonathan. II. Title.
JF1001.W37 2011
324.9--dc22

2011011624

ISBN 13: 978-1-138-27410-5 (pbk)
ISBN 13: 978-1-4094-1072-0 (hbk)

Contents

PART III ELECTORAL GEOGRAPHY IN THE US

List of Figures

List of Tables

List of Contributors

John Agnew, Department of Geography, University of California at Los Angeles, USA.

Thomas Chapman, Department of Political Science and Geography, Old Dominion University, USA.

Heather Hollen, Department of Geography, University of Oklahoma, USA.

Ron Johnston, School of Geographical Sciences, University of Bristol, UK.

Jonathan Leib, Department of Political Science and Geography, Old Dominion University, USA.

Daniel McGowin, Department of Humanities, Alabama State University, USA.

Toby Moore, Research Triangle Institute International, USA.

Erinn P. Nicley, Department of Geography, University of Illinois at Champaign-Urbana, USA.

Charles Pattie, Department of Geography, University of Sheffield, UK.

Nicholas Quinton, Department of Geography, Florida State University, USA.

Luis D. Sánchez-Ayala, Department of History, Universidad de los Andes, Colombia.

Fred M. Shelley, Department of Geography, University of Oklahoma, USA.

Michael Shin, Department of Geography, University of California at Los Angeles, USA.

Barney Warf, Department of Geography, University of Kansas, USA.

Gerald R. Webster, Department of Geography, University of Wyoming, USA.

PART I
Conceptual Foundations

Chapter 1
Introduction

Barney Warf and Jonathan Leib

Electoral geography – the analysis of spatial patterns in voting – has a long history and once possessed a distinguished status within the discipline of geography. From the 1960s through the 1980s, electoral geographers were important practitioners in the application of quantitative techniques and cartography to electoral data at the local and national scales. This body of work, largely empiricist in inspiration and focus, nonetheless generated a substantial literature that revealed the deeply spatial nature of elections, including electoral redistricting, economic and demographic factors, and the campaign strategies of parties and candidates (Taylor 1973; Taylor and Johnston 1979; Gudgin and Taylor 1979; Swauger 1980; Morrill 1981; Archer 1988; Archer and Shelley 1986, 1988; Archer et al. 1985; Archer and Taylor 1981; Johnston 1982; Johnston et al. 1990; Martis 1982; Martis 1989). Related works were concerned with the state, social relations, and the socio-spatial context of ideology, in which elections are seen as arenas in which subjects express their preferences within structural constraints ranging from the local scale to the world system (Agnew 1996; Flint 2001; Johnston and Pattie 2003). The vast majority of such projects focused on the US, with a few notable examples drawn from Europe (Agnew 1995; Johnston and Pattie 2006; Adams 2007). This literature has offered a rich, detailed portrait of the spatiality of elections at the national and local levels; redistricting and gerrymandering; shifts in voter preferences, turnout rates, and correlations with various socioeconomic variables; and neighborhood effects on political behavior.

Over the last three decades, however, electoral geography has fallen into senescence. As human geography became increasingly preoccupied with conceptual matters approached through various forms of social theory and political economy, electoral geographers' obsession with techniques and data, reflecting an unrepentant positivist or naïve empiricist outlook, and their studied neglect of issues of theory and social context, left the subdiscipline unable to contribute substantially to contemporary conceptual debates. Publications in this area, accordingly, declined in number and status, and electoral geography degenerated into its own "moribund backwater." As well, the practitioners within the field of electoral geography reflected an earlier time period within the discipline of geography. While the discipline has witnessed a partial closing of its gender inequality gap in recent decades (though not as fast and as thoroughly as we would like), the overwhelming majority of electoral geography studies are still authored by males.

More recently, however, a renaissance of sorts has begun to emerge in electoral geography (Johnston 2005; Shin and Agnew 2008; Warf 2009; Webster, Chapman and Leib 2010). Many political geographers, as well as some members in related disciplines such as political science, have sought to bring new perspectives to the field, countering the traditional positivism, engaging with conceptual debates, and avoiding the overly American bias that once plagued the topic. Methodological advances such as Geographic Information Systems have allowed electoral data to be analyzed in novel and intriguing ways. New theoretical perspectives from political economy to post-modernist and post-structuralist approaches have been added to the electoral geography literature, accentuating the role of multi-scalar explanations in the field, and injecting a concern for class, identity, ethnicity, gender, and sexuality within electoral geography studies. At the same time, when combined with more traditional spatial-analytic methodologies, these perspectives have provided greater insight through newer "mixed methods" approaches to the study of electoral geography. Moreover, the political environment itself has become more conducive to the subdiscipline's reconstruction, particularly in the US, where, since the heated presidential election of 2000, academic work on the spatiality of elections has intensified. Thus, theoretically, conceptually and pragmatically, electoral geography is poised to re-establish itself as an important part of the intersections between politics and geography.

This volume seeks to expedite the resurgence of electoral geography through a series of chapters that integrate new conceptual and theoretical approaches with a broad array of case studies from around the world. Key questions that tie the theoretical and empirical sections together include:

1. How has electoral geography changed since the 1980s, when the last wave of works in this subdiscipline appeared?
2. In what ways does contemporary scholarship in social theory inform the analysis of elections and their spatial patterns?
3. How has electoral geography been reconfigured by social and technological changes such as new requirements for voter identification?
4. As electoral geography distances itself from its previous obsession with the United States, how has the comparative analysis of elections informed the field?

Summary of Chapters

The chapters in this volume fall into three major categories. The first set deals conceptually with theoretical and conceptual topics, including the evolution of electoral geography, lines of debate, and silences and omissions. The second group consists of case studies drawn from the UK, Italy, Spain, and Taiwan. The third focuses on the United States; some are concerned with elections at the national scale, while others focus on more local contexts. These chapters also deal with different types of electoral geography topics, including general and primary candidate elections, referenda, voter eligibility issues, as well as the varied impacts of different types of voting and registration procedures and systems.

Jonathan Leib and Nicholas Quinton begin by tracking electoral geography's trajectory over the past 20 years. Analyzing 224 articles published over the last two decades, they trace major topic areas of electoral geography research, such as voting studies, electoral systems, redistricting, and pedagogic issues, as well as the diverse theoretical perspectives that have been brought to bear on the subject, such as spatial analysis, political economy, and poststructuralist interpretations. The latter, in particular, has led to a thorough reconceptualization of power within electoral geography and studied attempts to overcome the chasm that separated the analysis of elections from wider issues of class, gender, ethnicity, sexuality and inequality.

One of the signs of electoral geography's maturation is the growing number of case studies not located in the United States. Britain, of course, has long been the subject of attention from electoral geographers. Ron Johnston, one of human geography's most well known and prolific contributors, together with Charles Pattie use the British context to examine the impacts of different voting systems, including first-past-the-post, supplementary vote, and alternative vote, on national and local elections there, as well as comparisons with similar systems in other countries. Each of these approaches has strengths and weaknesses, as no voting system is perfect, but their adoption inevitably entails careful attention to the context and spatial specifics in which they are implemented.

In an indication of electoral geography's growing methodological maturity, Michael Shin and John Agnew use geographic information systems and geographically weighted regression (GWR) to explore the electoral geographies of Italy's *Lega Nord*, a secessionist and anti-immigrant group that has recently enjoyed rising popularity. Their case study reveals the necessity to take spatial structures into account when conducting statistical analysis of electoral patterns.

Erinn P. Nicley introduces a cultural political economy approach to examine the rise and fall of the regionalist political party, the *Bloque Nacionalista Galego* (BNG), in regional elections in Spain's northwest province of Galicia from 1985 to 2005. In his analysis, Nicley uses this cultural political economy perspective to bring political economy, cultural studies, and semiotics into the study of electoral geography, and, combined with GIS analysis, seeing these elections as moments in time that both are influenced by and influence changing political, economic and social life in northwestern Spain.

Traditionally, English-language studies of electoral geography have focused almost exclusively on elections in North America, Europe, and Australia/New Zealand. Daniel McGowin works to rectify this imbalance through his study of Taiwan/Republic of China's four direct presidential elections from 1996 to 2008. He traces the hegemony of the KMT, viewed by many Taiwanese as a foreign party that arrived from the mainland in the chaos of the 1940s. In the tussles between the KMT and rival parties in these elections, issues of place, identity, ideology and Taiwan's future status become hopelessly intertwined.

Puerto Rico, caught between its roles as US Commonwealth and an autonomous island with its own political character, is analyzed by Luis D. Sánchez-Ayala. In this chapter, Sánchez-Ayala focuses on the multiple attempts to resolve the island's final status through electoral referenda: whether as a Commonwealth, a US state, or an independent country. He focuses on a key issue within this debate over the island's future status: who would be considered eligible to vote in a future referendum. Would the electoral universe for a future referendum be limited to only those who live on the island? Would those who were born on the island but have moved to the mainland be included? Or would those born and raised on the mainland, but who can trace their ancestry back to the island, be also allowed to vote? Because there are more Puerto Ricans living on the US mainland than on the island itself, this issue is intertwined with complex questions about the construction of Puerto Rican identity/identities both on the island and the mainland.

In the US context, Barney Warf describes the impacts of class, ethnicity, and religion in the historic 2008 presidential election, which led to the first African-American president, Barack Obama. Contrary to stereotypes that portray Democrats as the party of the working class and Republicans as the party of the wealthy, he demonstrates that Obama's support tended to rise steadily with average household income level. Using GWR, he concludes that, holding other influences constant, many voters in each group tended to support Obama more strongly when they lived outside of regions in which their ethnicity or religion dominated, perhaps indicating a subtle geography of appreciation for difference.

Toby Moore summarizes the spatiality of several dimensions of American voting in the 2008 elections, including same day versus election day registration, early voting, provisional voting, voter identification laws, and voting technologies, all of which combine to shape local and national outcomes in complex ways, as well as emerging issues such as the impacts of votes cast by citizens located overseas.

While numerous studies exist, both historical and contemporary, of the geography of US presidential general elections, Fred M. Shelley and Heather Hollen examine an understudied yet vitally important aspect of these elections, the Presidential primary election process. Using the 2008 presidential primaries as a case study, Shelley and Hollen examine the important geographical aspects of the American primary election system that helped lead to the nomination of Barack Obama and John McCain by the country's two main political parties.

Using measures of spatial autocorrelation, Nicholas Quinton and Gerald R. Webster analyze spatial patterns of voting patterns in multiple elections in Alabama, both partisan contests and "culture war" referenda, demonstrating how quantitative approaches can be combined with contextually-sensitive studies in a "conciliatory" manner. As the culture wars flared across this deeply conservative state, they exposed deep and persistent geographic divides among the state's regions in which race, ideology, and place were inseparably fused.

Finally, Thomas Chapman analyzes the geographical contours of Florida's 2008 marriage amendment referendum, in which proponents sought to place a ban on same-sex marriage within the state's constitution. Taking a cue from the recent subgenre of work that invokes a critical and post-positivist approach towards electoral outcomes and the political geographies of sexuality, Chapman examines the electoral geographies of Florida's referendum results through the lens of the cultural politics of sexuality.

While diverse in numerous ways, the overall goal of the 11 chapters that follow is to outline the myriad theoretical, conceptual and methodological perspectives and applications that together are ushering in electoral geography's revitalization.

References

Adams, P. 2007. *Atlantic Reverberations: French Representations of an American Presidential Election.* London: Ashgate.

Agnew, J. 1995. The rhetoric of regionalism: The Northern League in Italian politics. *Transactions of the Institute of British Geographers* 20: 156–72.

Agnew, J. 1996. Mapping politics: How context counts in electoral geography. *Political Geography* 15: 129–46.

Archer, J. 1988. Macrogeographical versus microgeographical cleavages in American Presidential elections: 1940–1984. *Political Geography Quarterly* 7: 111–26.

Archer, J. and F. Shelley. 1986. *American Electoral Mosaics.* Washington, DC: Association of American Geographers.

Archer, J. and F. Shelley. 1988. The geography of US presidential elections. *Scientific American* 259: 44–53.

Archer, J., G. Murauskas and F. Shelley. 1985. Counties, states, sections, and parties in the 1984 presidential election. *Professional Geographer* 37: 279–87.

Archer, J. and P. Taylor. 1981. *Section and Party: A Political Geography of American Presidential Elections.* New York: John Wiley, Research Studies Press.

Gudgin, G. and P. Taylor. 1979. *Seats, Votes and the Spatial Organisation of Elections.* London: Pion.

Johnson, R. 1982. The changing geography of voting in the United States: 1946–1980. *Transactions of the Institute of British Geographers* 7: 187–204.

Johnston, R. 2005. Anglo-American electoral geography: Same roots and same goals, but different means and ends? *Professional Geographer* 57: 580–87.

Johnston, R. and C. Pattie. 2003. Representative democracy and electoral geography. In J. Agnew, K. Mitchell and G. Toal (eds) *A Companion to Political Geography.* Oxford: Blackwell, 337–55.

Johnston, R. and C. Pattie. 2006. *Putting Voters in Their Place: Geography and Elections in Great Britain.* New York: Oxford University Press.

Johnston, R., F. Shelley and P. Taylor (eds) 1990. *Developments in Electoral Geography*. London and New York: Routledge.
Martis, K. 1982. *The Historical Atlas of United States Congressional Districts: 1789–1983*. New York: The Free Press.
Martis, K. 1989. *The Historical Atlas of Political Parties in the United States Congress: 1789–1989*. New York: Macmillan.
Morrill, R. 1981. *Political Redistricting and Geographic Theory*. Washington, DC: Association of American Geographers.
Shin, M. and J. Agnew. 2008. *Berlusconi's Italy: Mapping Contemporary Italian Politics*. Philadelphia: Temple University Press.
Swauger, J. 1980. Regionalism in the 1976 presidential election. *Geographical Review* 70: 157–66.
Taylor, P. 1973. Some implications of the spatial organization of elections. *Transactions of the Institute of British Geographers* 60: 121–36.
Taylor, P. and R.J. Johnston. 1979. *Geography of Elections*. London: Penguin.
Warf, B. 2009. The US electoral college and spatial biases in voter power. *Annals of the Association of American Geographers* 99(1): 184–204.
Webster, G., T. Chapman and J. Leib. 2010. Sustaining the "societal and scriptural fence": Cultural, social, and political topographies of same-sex marriage in Alabama. *Professional Geographer* 62: 211–29.

Chapter 2
On the Shores of the "Moribund Backwater"?: Trends in Electoral Geography Research Since 1990

Jonathan Leib and Nicholas Quinton

In the 1970s and 1980s, political geography emerged from its "moribund backwater" (quoted in Berry 1969). In the forefront of this resurgence was the subfield of electoral geography, as many of the leading political geographers of the time helped to advance the subdiscipline (e.g., Taylor and Johnston 1979). The rise of electoral geography culminated in a 1988 conference in Los Angeles and the subsequent 1990 publication of papers from that conference in the book, *Developments in Electoral Geography*, edited by three prominent political geographers: Ron Johnston, Fred M. Shelley and Peter Taylor. However, criticisms of major strains of work in electoral geography were also on the rise at the time, with such charges of "methodological obsession" (Agnew 1990, 15), "rampant empiricism" (Shelley, Johnston and Taylor 1990, 1), and a lack of concern for social theory leveled against much of the work in electoral geography (see also Agnew 1987, Reynolds 1990).

The purpose of this chapter is to analyze recent trends in electoral geography research since the publication of *Developments in Electoral Geography* in 1990. First, we examine whether electoral geography has declined as an important subfield of political geography. Second, we assess trends in the topics studied and theoretical approaches employed within electoral geography. We conclude by assessing to what extent electoral geographers have responded to the criticisms and critiques of the subdiscipline issued 20 years ago.

To address these issues, we have created a database of 224 electoral geography journal articles published between 1990 and 2007. Before we discuss our survey of the subdiscipline through these articles, we need to address how we decided what to include and what not to include within this database.

First, we chose to examine electoral geography articles published in major geography journals, given the ability of databases that allow us to perform a fairly comprehensive search of these journals. We have included in our survey human geography journals indexed in the ISI-Journal Citation Reports database, a standard source for assessing major journals in the field. To this group, we have added additional non-ISI journals that we know have published a variety of electoral geography articles (such as the *Southeastern Geographer*). However, we recognize that it is nearly impossible to track down every non-ISI journal published that might contain electoral geography articles. As a result, we know we have missed some electoral geography articles published in non-ISI journals over the past 20 years (e.g., Ingalls and Strickland 1992). In addition, we have limited ourselves to English-language human geography journals. We recognize that, as a result, we have missed some electoral geography articles from journals published in other languages (e.g., Caviedes 1993). As well, our emphasis on journal articles means that we have not included books published in electoral geography, books that contain electoral geography analyses, chapters in edited volumes and essays in encyclopedias (for example, Agnew 2002b, Secor 2004, Leib and Webster 2009).

Has Electoral Geography Declined as an Important Subfield?

The first issue we address is whether electoral geography has declined as an important subfield within political geography. Our anecdotal observations gleaned from conferences and journals suggest that the field of electoral geography and the number of electoral geographers has been in decline over the past 20 years. As John Agnew (2003, 613) put it, by the late 1980s the focus on "studies of voting in electoral geography … seemed passé to many geographers." Writing in 2003, Agnew (2003, 613) suggested that while "much effort in political science and political sociology is still put into electoral studies … with a few notable exceptions, leading political geographers have moved on to other, presumably greater, things." It is telling that one of the leading current undergraduate political geography textbooks, Painter and Jeffrey's 2009 *Political Geography*, directs readers to *Developments in Electoral Geography* (Johnston, Shelley and Taylor 1990) "for a broader survey of the field of electoral geography" (Painter and Jeffrey 2009, 102), thereby suggesting the paucity of work conducted in this area of study over the past two decades.

So how has electoral geography fared? By our count, between 1990 and 2007 some 224 journal articles were published on topics that we considered to be electoral geography, or an average of about 12 per year (Figure 2.1). In terms of the trends over time, we see a spike in the number of electoral geography articles from 1995 to 2002. Though speculative, we attribute this spike to several factors. First are factors in terms of the publishing of the journals themselves, including guest edited issues of journals on electoral geography topics, as well as plenary lectures and commentaries (such as John Agnew's [1996] and Ron Johnston's [2002] in *Political Geography*), which increase the number of electoral geography articles in these years (though an increase in such articles and guest-edited issues may also be reflective of an increased interest in electoral geography). Secondly, electoral geography analyses of important political events of the mid- to late-1990s and early 2000s also helped lead to this spike: such as post-Cold War elections in former communist states, the US Supreme Court rulings of the mid- to late-1990s on the use of race in redistricting, and the disputed 2000 US presidential election.

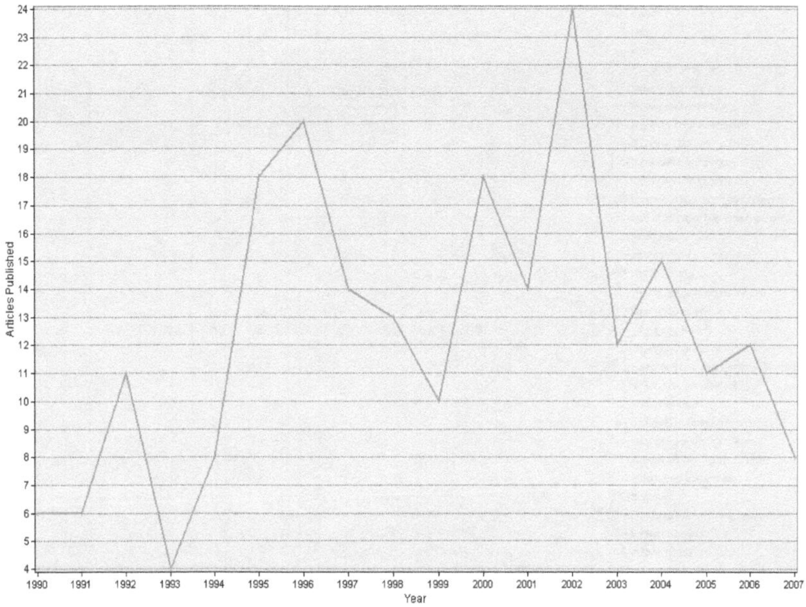

Figure 2.1 Number of Journal Articles on Electoral Geography Published, 1990–2007

In terms of journals, we found at least one electoral geography article in 28 different journals (Figure 2.2). Of these 28 journals, it is not surprising that about half of the 224 articles were published in one journal, *Political Geography,* the field's flagship journal. The only other two journals containing at least 5 percent of the electoral geography articles were *Post-Soviet/Eurasian Geography and Economics* (with 21 articles) and the *Southeastern Geographer* (with 13 articles). When we considered who was publishing electoral geography research, these 224 articles were published by 157 different authors (we assigned multiple authored articles to the lead author) from multiple disciplines. With that said, the disciplinary homes of these authors were overwhelmingly geography (59 percent) and political science (34 percent).

We should also note that the vast majority of geographers publishing articles in electoral geography during this time period were male. Only five female geographers published a single-authored/lead-authored article in the journals we surveyed from 1990 to 2007. Therefore, only 3.9 percent of all such electoral geography articles were single- or first-authored by a female geographer, a smaller percentage than that of female political geographers as a whole, and far below the 36.2 percent of all geographers who are women (as measured by 2009 membership figures from the Association of American Geographers) (AAG 2009).

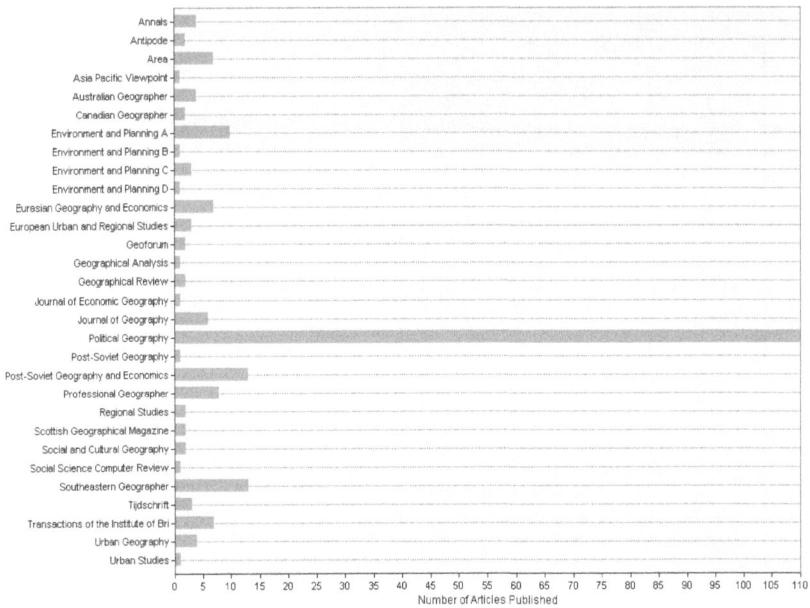

Figure 2.2 Number of Articles on Electoral Geography Published by Journal, 1990–2007

When we map out the sites of electoral geography research, the United States and the United Kingdom dominate: 39 percent of all articles focus on the US, whether at the local, state, regional or national levels, while 20 percent focus on the UK (Figure 2.3). This is probably not surprising given that we are looking at English-language journals and that many of the authors publishing in these journals are either from the US or the UK. Third, behind the US and the UK, are studies of elections in Russia, reflecting an interest in the electoral process there since the breakup of the USSR. In addition, starting with Martis, Kovacs, Kovacs, and Peter's 1992 article on elections in Hungary, we have seen an additional eight articles on elections in other former Soviet states or their former Eastern European satellites.

However, the lack of coverage of elections in the rest of the world is a bit startling. We counted only eight articles since 1990 dealing with elections in Asia (three of these were about Israeli elections), and only four articles on Latin America (with three of these on Mexico). Over this time period, we can only find three articles about elections in Africa. Certainly there is much room for English-language electoral geography research on the non-North American and European world.

As discussed above, about half of all electoral geography articles appear in the journal, *Political Geography*. The percentage of geographers and political scientists publishing electoral geography articles in *Political Geography* is similar to the percentage of articles published in journals as a whole: 62 percent

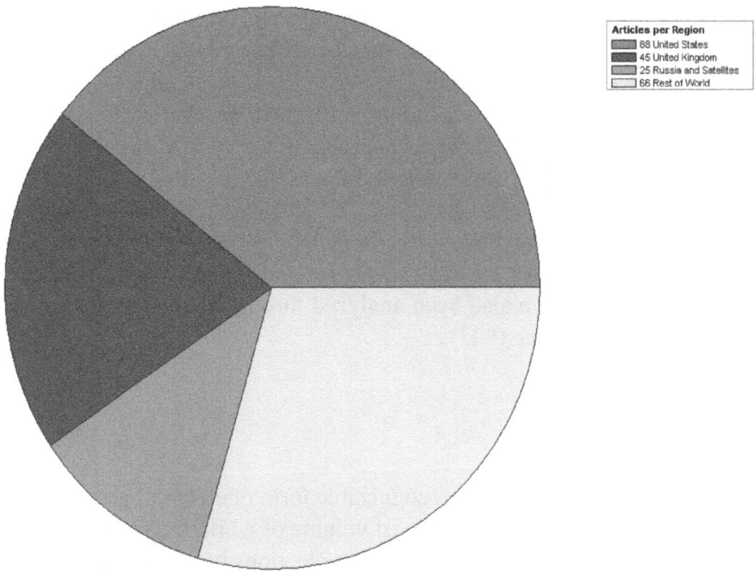

Figure 2.3 Country of Study in Journal Articles on Electoral Geography, 1990–2007

of articles in *Political Geography* are published by geographers, with 33 percent published by political scientists. We suggest there are two explanations for the relatively high percentage of political scientists publishing articles about elections in *Political Geography*: first, political scientists are still carrying out many studies of elections (and some are taking geographic perspectives more seriously [e.g., Gimpel, Karnes, McTague and Pearson-Merkowitz 2008]), and second, the journal *Political Geography* is highly ranked in the ISI Journal Citation Reports for political science. For 2009, the journal ranked sixth among 112 political science journals ranked by ISI. Hence for political scientists looking for a "top 10" publication, *Political Geography* is a good outlet.

Topics of Electoral Geography Research

What topics have been covered under the heading of electoral geography over the past 20 years? While the obvious answer is elections, the specific topics covered have been far ranging: from analyzing spatial dimensions of post-Soviet Russian election outcomes, to theoretical and empirical studies of redistricting and gerrymandering, to suggesting methods for teaching about the American Electoral College in middle- and high-school classrooms. Given this range, we have grouped electoral geography studies into four broad categories which we discuss below in more detail: vote studies, comparisons amongst and critiques of electoral systems, discussions and examinations of the (re)districting process, and methods to bring electoral studies into the classroom.

Before we go further, we should also mention that analysis in electoral geography is not only carried out on a number of topics, but also across a variety of political scales, including supranational (e.g., Laponce 2004), national (e.g., Shin 2001), and subnational (e.g., Chapman, Leib and Webster 2007). As well, electoral geography studies are carried out over a variety of electoral contexts. While most involve popular votes for candidates, parties, and issues, voting behavior within elected legislatures have also been analyzed through the study of legislative roll call voting (e.g., Webster 1992).

Vote Studies

Vote studies are likely the most recognizable form of electoral geography, and this is the category that includes the largest volume of work. Dissecting the intricacies of popular votes in a particular election or elections has been a common practice for electoral geographers since the 1950s and before.

These studies focused on the votes cast for a candidate, votes cast for a party, those on single-issue/referendum, or roll-call legislative votes. In general, vote studies examined either the spatial distribution of compositional factors (such as race or income) associated with electoral outcomes (Clem and Craumer 1995a, 1995b, 1996, 2000 for examples, or King 1996 for advocacy of this type of work), or place-based contextual factors (Agnew 1994, 1995, 1996; Rasmussen 2006; Shin 2001; Shin and Agnew 2002) and their relationship to voting behavior.

Further, vote studies, like electoral geography as a whole, have covered a variety of topics. For example, there have been studies that highlighted regional polarization within states or growing sentiments of nationalism and how that related to votes cast (Agnew 1995; Giordano 1999, 2001a, 2001b). Other studies have highlighted the difference between urban and rural voters (Clem and Craumer 1995b), or central city and suburban voters (Hodge and Staeheli 1992; Ley 1994). Still other geographers have studied turnout (Kohfeld and Sprague 2002; Lee and Brunn 1996; Pattie and Johnston 1998), tied economic factors to voting behavior (O'Loughlin, Shin, and Talbot 1996; Pattie, Dorling, and Johnston 1997; West 2005), and explored the effects of elections on migration (Lo and Teixeira 1998). This type of work has also included a series of studies that related debates over identity including race/ethnicity, gender, and sexual orientation to electoral outcomes (Brown, Knopp, and Morrill 2005; Chapman, Leib, and Webster 2007; O'Reilly and Webster 1998; Webster 2000b). Thus, vote studies cover a number of disparate topics in different ways, but all of them offered analysis of a particular vote or series of votes.

Electoral Systems

Electoral systems, or perhaps more appropriately voting systems, are critical to election outcomes. Voting systems are the framework through which a person's vote is transferred to the candidate, party, or issue of their choice. Thus, understanding vote systems is important for electoral geography.

Electoral system research has approached the issue of voting systems from several directions. One group of articles assessed how effective a voting system is at translating votes into representation. So, for example, there have been studies comparing various systems in an effort to determine which provided the highest quality of representation, such as Arrington and Ingalls (1998), who offered an example of this approach with an analysis of six voting systems at use in the US based upon nine measures of feasibility, and Johnston (2007), who sought to outline Australian exceptionalism in terms of voting practices in a similar broad based study of voting systems. In a slightly different approach, some authors have assessed bias between the proportion of votes cast and the proportion of seats awarded. Johnston et al. (2002) explored this issue in Great Britain with attention to shifting bias from the Conservatives to the Labor party. Yet another related topic was offered by Johnston et al. (2006) in their attempt to replay the 2000 and 2004 US presidential elections under the Maine/Nebraska Electoral College system (in these two states, rather than electoral votes being awarded on a "winner-take-all" basis, their electoral votes are allocated based both on a "winner-take-all" basis and by congressional district). So, the effectiveness of representation has been one avenue of research about electoral systems.

Another avenue of research dealt with legitimacy. Vote systems must attain a level of legitimacy within an electorate in order to gain popular acceptance and be implemented (Arrington and Ingalls 1998). Included in our cross-section of vote systems are those studies that focused on implementation, including the study of public debates over changing the electoral systems in Israel (Waterman and Zefadia 1992). In a similar vein of research, Hazan (1999) followed the process of introducing primaries into the 14th Israeli Knesset (1996–1999), their ramifications and geographical significance. But some authors have gone beyond implementation to assess the legitimacy of the mechanics of the voting process itself. These authors have studied the means through which votes are initially cast and are then counted. Much of this work occurred in the aftermath of the disputed 2000 US presidential election, including such topics as residual votes in Florida (Warf 2006b), in terms of impact on the minority community (Warf 2006a), and in reference to vote tally irregularities associated with voting technology (Leib and Dittmer 2002).

(Re)Districting

Topics of vote studies and electoral systems are closely related to that of (re) districting. Redistricting, the "process of drawing electoral district boundaries for the purpose of electing members to a legislative body" (Leib 2006, 404), has been a staple of electoral geography research going back as far as the early work of Carl Sauer (1918). Though most associated with US redistricting, overviews of the redistricting process have not been limited to studies in the US. Broad overviews have been published on the districting process in the UK (see Johnston 1992), India (see Singh 2000), and Canada (see Belanger 2003).

In the US, such studies have been important in terms of examining the partisan impact of redistricting, the impact of redistricting for effective representation of racial and ethnic minority groups, and ways of creating and measuring fair and effective redistricting plans. Given the role of the courts in the redistricting process over the past 20 years, a number of these studies have examined the impact of court decisions and case law on the political-geographic division of space. Of these, a series of articles have provided broad overviews of the precedents established in case law and how recent court rulings have impacted existing ways of regulating, implementing and evaluating the redistricting process (e.g., Forest 1995; Forest 1996; Forest 2004b; Leib 1998; Webster 1997, 2000a, 2002). Other US-based studies have examined specific dimensions of the redistricting process, such as gerrymandering (e.g., Ingalls and Moore 1995; Lennertz 2000) and malapportionment (Rallings, Johnston, and Thrasher 2004).

More than in other areas of electoral geography research, redistricting studies have been produced by a wide array of authors, as authors from across the discipline of geography, outside of geography, and outside of academe have contributed. Along with political geographers and political scientists, redistricting issues attract GIS experts, statisticians, Department of Justice professionals, and social theorists alike (though, of course, there is the potential for overlap among these categories). Such studies have included the use of quantitative applications to meet legal standards for (re)districting (Cirincione, Darling, and O'Rourke 2000; Eagles, Katz, and Mark 1999; Thill 1998), such as an evaluation of compactness (Belanger and Eagles 2001; Flaherty and Crumplin 1992; Horn, Hampton, and Vandenberg 1993), and the use of Bayesian statistics to detect electoral manipulation (Altman 2002). In addition, there are (re)districting studies that offered a critical perspective, such as affirmative (re)districting (Webster 1993) in establishing/maintaining voting rights (Webster 1995, 2007). And, in a similar vein, post-structuralist ideas of hybridity and multiplicity have emerged in research discussing communities of interest (e.g., Forest 2004a; Leib 1998). Thus, the body of work on (re)districting is produced by a unique web of authors within and outside of electoral geography.

Education/Classroom Methods

Finally, a small, but critical, component of electoral geography relates to praxis in education. We have several articles that have been published in geography journals that provide approaches to relate issues of elections and representation to students in a classroom setting. Depoali (1993) offered various strategies to aide in teaching middle- and high-school students about the American Electoral College. Jones (1997) highlighted geographical perspectives on a range of issues that include delineating electoral districts in an effort to help inform public policy debate amongst students. Finally, Webster (2004) provided examples of the (re) districting process to help incorporate social justice issues into the classroom.

Theoretical Perspectives within Electoral Geography Research

Topics studied within electoral geography over the past 20 years have been disparate to say the least, ranging from poll taxes in Scotland (McCormick 1996) to grassroots movements in India (Currie 1998). Each study, however, fits within at least one of the three major theoretical approaches, as outlined by John Agnew (2002a), which have dominated political geography in its post-1970 renaissance. These theoretical frameworks are the spatial-analytic tradition, political economy, and post-structural approaches. We outline these below.

Spatial Analysis

The first and largest set of electoral geography studies are those that are firmly situated within the spatial-analytic tradition. Spatial analysis is typically conceptualized as the application of statistical methodologies to quantifiable data, though this is not necessarily always the case (Agnew 2002a). However, the dominant manifestation from this theoretical outlook in electoral geography has undoubtedly been quantitative in nature. The spatial analysis tradition dominated electoral geography in the 1970s. This is not surprising, in that the rise of electoral geography in that decade, leading the way in the resurgence of political geography, coincided with the high point of spatial analytic methodologies within human geography. Elections provided a "natural" area of study for geographers within this tradition, as elections provide vast amounts of quantitative data (i.e., votes) that can be mapped, manipulated and analyzed through sophisticated statistical techniques.

Since 1990, electoral geography work from a spatial-analytic perspective has included analyses of the channels of information that political discussion travels, including friends and neighbors effects (Johnston, Jones, Sarker, Burgess, and Bolster 2004; Kramer 1990; Rallings, Johnston, and Thrasher 2004; Sui and Hugill 2002). There have also been explorations of a variety of political cleavages that divide selected electorates under the spatial-analytic umbrella (Hodge and Staeheli 1992; Perepechko, Kolossov, and ZumBrunnen 2007). One study has been published expounding on the cube rule in search of fractality (Maloney, Pearson, and Pickering 2003). Further, there are studies that have sought to deepen the level of quantitative sophistication in terms of eigenfunction analysis (Thill 1998), combating the ecological fallacy (Johnston and Pattie 2003; Owen and Grofman

1997), using Bayesian statistics to detect electoral manipulation (Altman 2002), evaluating the significance of temporal and spatial data (Fernandez-Duran, Poire, and Rojas-Nandayapa 2004), and forwarding the use of local area statistics (LISA) in electoral analysis (Darmofal 2006).

Finally, there are a series of studies that employ statistical methodologies to describe electoral outcomes. Some catalog demographic trends as they vary through space over particular elections or groups of elections (Archer 2002; Clem and Craumer 1995a; Clem and Craumer 1995b). Others include discussions of normal and deviating elections as parts of broader electoral trends (Forrest, Johnston, and Pattie 1999). These studies have also been extended into discussions of the mechanics of large scale electoral changes, including flow-of-vote as opposed to the uniform swing hypothesis (see Pattie and Johnston 1993).

Political Economy

Over the past 20 years, political economy approaches have been less popular within electoral geography than studies those from a spatial analytic perspective. There are at least three notable studies that incorporated political economy perspectives during this time period (though having said this, there were electoral geography studies from a political economy approach in the 1980s, when political economy perspectives were more in favor within political geography (Oseikwame and Taylor 1984). First, Duncan (1991) explored the geography of support and power in Fijian elections. Next, Flint (2001) employed a world systems approach to explain the rise of the Nazi party in Germany. Finally, Varsanyi (2005) delves into the intersection between labor unions, undocumented immigrants, and political control. Despite these studies, it is apparent that this perspective has received less attention within electoral geography than it has within political geography in general over the past two decades.

Post-Structuralism

Lastly, there are studies from post-structuralist viewpoints. These have grown in popularity in recent years and now include studies utilizing feminist perspectives and those explicitly dealing with the politics of identity. Pieces of note include Webster and Leib's (2001) study of the display and meaning of the Confederate flag in South Carolina with consideration to the contestation of identity and "place." Forest's (2004b) study of racial identity in *Shaw v. Reno* and *Easley v. Cromartie* is another example. The formation of regional political identities in Mexico from a feminist perspective also illustrates the point (Nelson 2006). Finally, Rasmussen (2006) offers a critique of the 2000 Nebraska anti-gay marriage referenda in terms of creating a decidedly non-gay "place" known as Nebraska.

Answering the Call?: Electoral Geography Research since 1990

To what extent, then, have electoral geographers answered the call for change made in *Developments in Electoral Geography*? We cannot provide an answer without contextualizing the initial call. Accompanying the growing critiques of electoral geography in *Developments* and those criticisms from other sources within the late 1980s and early 1990s was the growing engagement with post-structural perspectives across geography. Though the critiques levied against electoral geography were not necessarily made from explicitly post-structural perspectives, the critiques themselves can not be separated from the social milieu of academic geography at that juncture in time and space. Discipline-wide criticisms included feminist and post-colonial interjections into the meta-narratives produced by both positivist approaches and critical strands of geography. These post-structural critiques were followed by others concerned with identity, including queer theory, and post-structuralisms detailing heterogeneous associations, such as actor-network theory. At present the uses of post-structuralist ideas have become a significant trajectory of geographic inquiry in their own right, and are no longer simply challenging positivist epistemologies at each turn.

Despite criticism, electoral geographers did participate in these broader changes within academic geography. A critical component of geographic post-structuralisms is that space and place is produced and power relations are central to its production. In the sense that power is critical to spatial relations, each of the political economy studies embodies an answer to the call for change in electoral geographies. Spaces of capital accumulation are at the forefront of the political debates included in these studies and each involved the exercise of power throughout. Another body of work focused on power relations and space and place involved in redistricting(s) in the United States. These studies focused on redistricting designed to provide fair and effective representation for African-American citizens. Thus, issues of political representation based upon racial and ethnic identity were drawn to the forefront of these debates concerning issues of delineating spaces of representation.

Another component of post-structural geographies is the multiplicity of channels through which power travels. Thus, some of the most recognizable examples of changing epistemologies manifested themselves in the studies of identity and electoral politics listed above. Race, gender, and sexuality were all brought into focus as contestations of spatialized identities as played out within electoral processes. The diversity in these cleavages of power and their usage to produce spaces and places leaves little doubt that post-structuralisms have influenced the current shape of electoral geography.

Electoral geography studies that dealt with political economy, redistricting, and identity politics each engaged another axis by which post-structural geographies can be differentiated; contextualized and contingent power relations. The most sustained engagement with this particular idea has been found in the work of John Agnew (e.g., 1987, 1996, 2002b). Through the context of peoples' lives these works on place and its political implications have highlighted the contingent results of electoral processes. In the same sense that the outcomes of political debates over identity are not preordained in electoral processes, the outcomes of place-based political debates are contingent on the application of power in varied and unpredictable ways.

Surprisingly (or possibly not surprising), however, the most numerous engagements with post-structural ideas by electoral geographers has been in conjunction with the spatial analytic tradition. While post-structuralist studies throughout the broader discipline have by and large jettisoned quantitative methods, electoral geographers' engagements with post-structural ideas have embraced quantitative methods. Recent works have taken the relationship between the post-structural ideas of power, difference, and contingency and coupled them with quantitative methodologies. In particular recent work by John Agnew and Michael Shin (Shin and Agnew 2002; Shin and Agnew 2008), Richard Morrill, Larry Knopp, and Michael Brown (Brown, Knopp, and Morrill 2005; Morrill, Knopp, and Brown 2007), as well as Gerald R. Webster, Jonathan Leib, Thomas Chapman and Nicholas Quinton (Chapman, Leib, and Webster 2007; Webster, Chapman, and Leib 2010; Webster and Leib 2001; Webster and Leib 2002; Webster and Quinton 2010), have made inroads in coupling post-structural and quantitative geographies. All have deployed contexts and the capillaries that power traveled through within these places alongside quantitative measures. Thus, in these ways, attempts have been made to fuse post-structural ideas with the dominant positivist forms of geographic inquiry in electoral geography, quantitative methodologies.

Thus, has the call for change in electoral geography issued in 1990, that the field address its rampant empiricism, methodological obsession, and lack of concern for social theory, been answered? The answer appears to be that it has, at least to a limited extent. Though any claims to significant change must be tempered with the broader findings of this chapter. In actuality, the number of articles employing post-structural viewpoints has been small compared to the total number of articles we count as electoral geography. The paucity of engagement with post-structural approaches should not be surprising given the penchant for empiricism and quantitative methods shown by electoral geographers. Indeed, the overwhelming number of articles we identified that fit into the spatial analytic tradition shows that electoral geographers have by and large not answered the call for change, post-structural or otherwise. Still, approaches have changed and post-structural ideas have been employed in some accounts of electoral geographies, even though they have occasionally come in unlikely shapes.

To what extent, then, has electoral geography declined as an important subfield of political geography? The uses of post-structural ideas have undoubtedly been a part of the changes in the broader practice of political geography over the 20-year time period since the 1990 publication of *Developments in Electoral Geography*. At the same time, the maintenance of the quantitative tradition in electoral geography has distanced it from recent trends in political geography more generally. The answer then is mixed. As Murdoch (2006, 2) notes, "post-structuralism's endless variety means it becomes hard to draw any clear line or boundary around post-structuralist theory or its influence." In this sense we can say clearly that while on the one hand electoral geography has contributed to the discussion of post-structuralism more generally (though in its own way) and thus remains a relevant subfield of political geography, on the other hand further engagement with emerging trends within the discipline are necessary for it to regain its former prominent place within political geography.

References

AAG (Association of American Geographers) 2009. Profile of the AAG Membership, 2009. Available at http://www.aag.org/galleries/project-programs-files/2009_Membershipprofile.pdf [last accessed: October 27, 2010].

Agnew, J.A. 1987. *Place and Politics: The Geographical Mediation of State and Society*. Boston: Allen & Unwin.

Agnew, J.A. 1990. "From political methodology to geographical social theory? A critical review of electoral geography, 1960–87." In R.J. Johnston, F.M. Shelley, and P.J. Taylor (eds), *Developments in Electoral Geography*. London: Routledge, 15–21.

Agnew, J.A. 1994. "The national versus the contextual – The controversy over measuring electoral change in Italy using Goodman Flow-of-Vote estimates." *Political Geography* 13: 245–54.

Agnew, J.A. 1995. "The rhetoric of regionalism – The Northern League in Italian politics, 1983–94." *Transactions of the Institute of British Geographers* 20: 156–72.

Agnew, J.A. 1996. "Mapping politics: How context counts in electoral geography." *Political Geography* 15: 129–46.

Agnew, J.A. 2002a. *Making Political Geography*. London: Arnold.

Agnew, J.A. 2002b. *Place and Politics in Modern Italy*. Chicago: University of Chicago Press.

Agnew, J.A. 2003. Classics in Human Geography Revisited: Agnew, J.A. 1987: *Place and Politics*: *The Geographic Mediation of State and Society*. Author's Response. *Progress in Human Geography* 27: 609–14.

Altman, M. 2002. "Discussion – A Bayesian approach to detecting electoral manipulation." *Political Geography* 21: 39–48.

Archer, J.C. 2002. "The geography of an interminable election: *Bush v. Gore*, 2000." *Political Geography* 21: 71–7.

Arrington, T.S. and G.L. Ingalls. 1998. "The limited vote alternative to affirmative districting." *Political Geography* 17: 701–28.

Belanger, P. 2003. "Commissioned ridings: Designing Canada's electoral districts." *Political Geography* 22: 121–3.

Belanger, P. and M. Eagles. 2001. "The compactness of federal electoral districts in Canada in the 1980s and 1990s: An exploratory analysis." *Canadian Geographer-Geographe Canadien* 45: 450–60.

Berry, B.J. 1969. Review of *International Regions and the International System* (by B.M. Russett). *Geographical Review* 59: 450–51.

Brown, M., L. Knopp and R. Morrill 2005. "The culture wars and urban electoral politics: Sexuality, race, and class in Tacoma, Washington." *Political Geography* 24: 267–91.

Caviedes, C.N. 1993. "Estudios sobre la variabilidad espacial del voto en America Latina." *Inter-American Review of Bibliography* 43: 257–67.

Chapman, T., J.I. Leib and G.R. Webster 2007. "Race, the creative class, and political geographies of same sex marriage in Georgia." *Southeastern Geographer* 47: 27–54.

Cirincione, C., T.A. Darling, and T.G. O'Rourke. 2000. "Assessing South Carolina's 1990s congressional districting." *Political Geography* 19: 189–211.

Clem, R.S. and P.R. Craumer 1995a. "The geography of the Russian 1995 parliamentary election: Continuity, change, and correlates." *Post-Soviet Geography* 36: 587–616.

Clem, R.S. and P.R. Craumer 1995b. "Rayon-level analysis of the Russian election and constitutional plebiscite of December 1993." *Post-Soviet Geography* 36: 459–75.

Clem, R.S. and P.R. Craumer 1996. "Roadmap to victory: Boris Yel'tsin and the Russian presidential elections of 1996." *Post-Soviet Geography and Economics* 37: 335–54.

Clem, R.S. and P.R. Craumer 2000. "Spatial patterns of political choice in the post-Yel'tsin era: The electoral geography of Russia's 2000 presidential election." *Post-Soviet Geography and Economic* 41, 465–82.

Currie, B. 1998. "Public action and its limits: Re-examining the politics of hunger alleviation in eastern India." *Third World Quarterly* 19: 873–92.

Darmofal, D. 2006. "The political geography of macro-level turnout in American political development." *Political Geography* 25: 123–50.

Depoali, K. 1993. "The Electoral-College – not just every 4 years – some exercises in political-geography." *Journal of Geography* 92: 180–93.

Duncan, C.J. 1991. "Ethnicity, election and emergency – The 1987 Fiji general-election in the context of contemporary political-geography." *Political Geography Quarterly* 10: 221–39.

Eagles, M., R.S. Katz, and D. Mark. 1999. "GIS and redistricting – Emergent technologies, social geography, and political sensibilities." *Social Science Computer Review* 17: 5–9.

Fernandez-Duran, J.J., A. Poire, and L. Rojas-Nandayapa. 2004. "Spatial and temporal effects in Mexican direct elections for the chamber of deputies." *Political Geography* 23: 529–48.

Flaherty, M.S. and W.W. Crumplin. 1992. "Compactness and electoral boundary adjustment – An assessment of alternative measures." *Canadian Geographer-Geographe Canadien* 36: 159–71.

Flint, C. 2001. "A TimeSpace for electoral geography: Economic restructuring, political agency and the rise of the Nazi party." *Political Geography* 20: 301–29.

Forest, B. 1995. "Taming race – The role of space in voting-rights litigation." *Urban Geography* 16: 98–111.

Forest, B. 1996. "Regionalism in election district jurisprudence." *Urban Geography* 17: 572–8.

Forest, B. 2004a. "Information sovereignty and GIS: The evolution of "communities of interest" in political redistricting." *Political Geography* 23: 425–51.

Forest, B. 2004b. "The legal (de)construction of geography: Race and political community in Supreme Court redistricting decisions." *Social & Cultural Geography* 5: 55–73.

Forrest, J., R.J. Johnston, and C.J. Pattie. 1999. "The effectiveness of constituency campaign spending in Australian state elections during times of electoral volatility: The New South Wales case, 1988–95." *Environment and Planning A* 31: 1119–28.

Gimpel, J.G., K.A. Karnes, J. McTague, and S. Pearson-Merkowitz. 2008. "Distance-decay in the political geography of friends-and-neighbors voting." *Political Geography* 27: 231–52.

Giordano, B. 1999. "A place called Padania? The Lega Nord and the political representation of Northern Italy." *European Urban and Regional Studies* 6: 215–30.

Giordano, B. 2001a. "The contrasting geographies of 'Padania': The case of the Lega Nord in Northern Italy." *Area* 33: 27–37.

Giordano, B. 2001b. "'Institutional thickness', political sub-culture and the resurgence of (the 'new') regionalism in Italy – A case study of the Northern League in the province of Varese." *Transactions of the Institute of British Geographers* 26: 25–41.

Hazan, R.Y. 1999. "Constituency interests without constituencies: The geographical impact of candidate selection on party organization and legislative behavior in the 14th Israeli Knesset, 1996–99." *Political Geography* 18: 791–811.

Hodge, D.C. and L.A. Staeheli. 1992. "Social-relations and geographic patterns of urban electoral-behavior." *Urban Geography* 13: 307–33.

Horn, D.L., C.R. Hampton, and A.J. Vandenberg. 1993. "Practical application of district compactness." *Political Geography* 12: 103–20.

Ingalls, G.L. and J.L. Strickland. 1992. "No blurred edges; No crowded middle: Votes for Jesse Helms in 1984 and 1990." *North Carolina Geographer* 1: 46–54.

Ingalls, G.L. and T. Moore. 1995. "The present and future of racial gerrymandering: Evidence from North Carolina's 12th Congressional District." *Southeastern Geographer* 35: 58-74.

Johnston, R.J. 2002. "Manipulating maps and winning elections: Measuring the impact of malapportionment and gerrymandering." *Political Geography* 21: 1–32.

Johnston, R.J., F.M. Shelley, and P.J. Taylor, eds. 1990. *Developments in Electoral Geography*. London: Routledge.

Johnston, R.J. 2007. "Understanding Australian exceptionalism with electoral systems-and bringing geography in." *Australian Geographer* 38: 367–74.

Johnston, R.J. 1992. "Political-Geography of Contemporary Events: Redistricting in England Revisited." *Political Geography* 11: 579–87.

Johnston, R.J., K. Jones, R. Sarker, S. Burgess, and A. Bolster. 2004. "Party support and the neighbourhood effect: Spatial polarisation of the British electorate, 1991–2001." *Political Geography* 23: 367–402.

Johnston, R.J. and C. Pattie. 2003. "Evaluating an entropy-maximizing solution to the ecological inference problem: Split-ticket voting in New Zealand, 1999." *Geographical Analysis* 35: 1–23.

Johnston, R.J., C. Rallings, and M. Thrasher. 2002. "Electoral success, electoral bias, and Labour hegemony: Electoral system effects in English metropolitan boroughs." *Environment and Planning A* 34: 1303–17.

Johnston, R.J., D. Rossiter, and C. Pattie. 2006. "Changing the scale and changing the result: Evaluating the impact of an electoral reform on the 2000 and 2004 US Presidential elections." *Political Geography* 25: 557–69.

Jones, D.D. 1997. "The geography for life standards and American political geography." *Journal of Geography* 96: 55–60.

King, G. 1996. "Why context should not count." *Political Geography* 15: 159–64.

Kohfeld, C.W. and J. Sprague 2002. "Race, space, and turnout." *Political Geography* 21: 175–93.

Kramer, D.C. 1990. "Those people across the water just don't know our problems – An analysis of friends and neighbors voting in a geographically-split legislative district." *Political Geography Quarterly* 9: 189–96.

Laponce, J.A. 2004. "Turning votes into territories: Boundary referendums in theory and practice." *Political Geography* 23: 169–83.

Lee, D.O. and S.D. Brunn 1996. "Politics and regions in Korea: An analysis of the recent presidential election." *Political Geography* 15: 99–119.

Leib, J.I. 2006. "Redistricting." In B. Warf, ed., *Encyclopedia of Human Geography*. Thousand Oaks, CA: Sage, 404–5

Leib, J. I. 1998. "Communities of interest and minority districting after *Miller v. Johnson*." *Political Geography* 17: 683–99.

Leib, J. I. and J. Dittmer. 2002. "Florida's residual votes, voting technology, and the 2000 election." *Political Geography* 21: 91–8.

Leib, J.I. and G.R. Webster. 2009. "Electoral Districts." In R. Kitchin and N. Thrift, eds., *International Encyclopedia of Human Geography, Volume 3*. Oxford: Elsevier, 399–404.

Lennertz, J.E. 2000. "Back in their proper place: Racial gerrymandering in Georgia." *Political Geography* 19: 163–88.

Ley, D. 1994. "Gentrification and the politics of the new middle-class." *Environment and Planning D-Society & Space* 12: 53–74.

Lo, L. and C. Teixeira 1998. "If Quebec goes ... the 'exodus' impact?" *Professional Geographer* 50: 481–98.

Maloney, J., B. Pearson, and A. Pickering. 2003. "Behind the cube rule: Implications of and evidence against a fractal electoral geography." *Environment and Planning A* 35: 1405–14.

Martis, K., Z. Kovacs, D. Kovacs, and S. Peter. 1992. "The geography of the 1990 Hungarian parliamentary elections." *Political Geography* 11: 283–305.

McCormick, J. 1996. "Taxes and turnouts: The poll tax and the Scottish regional elections." *Scottish Geographical Magazine* 112: 147–57.

Morrill, R., L. Knopp and M. Brown 2007. "Anomalies in red and blue: Exceptionalism in American electoral geography." *Political Geography* 26 525–553.

Murdoch, J. 2006. *Post-structuralist Geography: A Guide to Relational Space.* London; Thousand Oaks, CA: Sage.

Nelson, L.K. 2006. "Geographies of state power, protest, and women's political identity formation in Michoacan, Mexico." *Annals of the Association of American Geographers* 96: 366–89.

O'Loughlin, J., M. Shin and P. Talbot 1996. "Political geographies and cleavages in the Russian parliamentary elections." *Post-Soviet Geography and Economics* 37: 355–85.

O'Reilly, K. and G.R. Webster 1998. "A sociodemographic and partisan analysis of voting in three anti-gay rights referenda in Oregon." *Professional Geographer* 50: 498–515.

Oseikwame, P. and P.J. Taylor 1984. "A Politics of Failure – the Political-Geography of Ghanaian Elections, 1954–1979." *Annals of the Association of American Geographers* 74: 574–89.

Owen, G. and B. Grofman. 1997. "Estimating the likelihood of fallacious ecological inference: linear ecological regression in the presence of context effects." *Political Geography* 16: 675–90.

Painter, J. and A. Jeffrey. 2009. *Political Geography* (Second Edition). London: Sage.

Pattie, C., D. Dorling and R. Johnston 1997. "The electoral geography of recession: Local economic conditions, public perceptions and the economic vote in the 1992 British general election." *Transactions of the Institute of British Geographers* 22: 147–61.

Pattie, C.J. and R.J. Johnston 1998. "Voter turnout and constituency marginality: geography and rational choice." *Area* 30: 38–48.

Perepechko, A.S., V.A. Kolossov, and C. ZumBrunnen. 2007. "Remeasuring and rethinking social cleavages in Russia: Continuity and changes in electoral geography 1917–1995." *Political Geography* 26: 179–208.

Rallings, C., R. Johnston, and M. Thrasher. 2004. "Equalising votes but enabling bias: The electoral impact of the 1977 and 1999 ward boundary reviews in London." *Urban Studies* 41: 1367–93.
Rasmussen, C.E. 2006. "We're no metrosexuals: Identity, place and sexuality in the struggle over gay marriage." *Social & Cultural Geography* 7: 807–25.
Reynolds, D.R. 1990. "Whither electoral geography? A critique." In R.J. Johnston, F.M. Shelley, and P.J. Taylor (eds), *Developments in Electoral Geography*. London: Routledge, 22–35.
Sauer, C.O. 1918. "Geography and the *Gerrymander*." *American Political Science Review* 12: 403–26
Secor, A.J. 2004. "Feminizing electoral geography." In L.A. Staeheli, E. Kofman, and L.J. Peake (eds) *Mapping Women, Mapping Politics: Feminist Perspectives on Political Geography*. New York: Routledge, 261–72.
Shelley, F.M., R.J. Johnston, and P.J. Taylor 1990. "Developments in electoral geography." In R.J. Johnston, F.M. Shelley, and P.J. Taylor (eds), *Developments in Electoral Geography*. London: Routledge, 1–11.
Shin, M.E. 2001. "The politicization of place in Italy." *Political Geography* 20: 331–52.
Shin, M.E. and J. Agnew 2002. "The geography of party replacement in Italy, 1987–1996." *Political Geography* 21: 221–42.
Shin, M.E. and J.A. Agnew. 2008. *Berlusconi's Italy: Mapping Contemporary Italian Politics*. Philadelphia: Temple University Press.
Singh, C.P. 2000. "A century of constituency delimitation in India." *Political Geography* 19: 517–32.
Sui, D.Z. and P.J. Hugill. 2002. "A GIS-based spatial analysis on neighborhood effects and voter turn-out: A case study in College Station, Texas." *Political Geography* 21: 159–73.
Taylor, P.J. and R.J. Johnston. 1979. *Geography of Elections*. London: Harmondsworth.
Thill, J.C. 1998. "A note on the matrix formulation of gerrymanders." *Environment and Planning B-Planning & Design* 25: 495–505.
Varsanyi, M.W. 2005. "The paradox of contemporary immigrant political mobilization: Organized labor, undocumented migrants, and electoral participation in Los Angeles." *Antipode* 37: 775–95.
Warf, B. 2006a. "Do voting technologies discriminate across the urban hierarchy? Evidence from two presidential elections." *Urban Geography* 27: 664–77.
Warf, B. 2006b. "Voting technologies and residual ballots in the 2000 and 2004 presidential elections." *Political Geography* 25: 530–56.
Waterman, S. and E. Zefadia. 1992. "Political-geography in practice II: Israeli electoral reforms in action." *Political Geography* 11: 563–78.
Webster, G.R. 1992. "Geography of a Senate confirmation vote." *Geographical Review* 82: 154–65.
Webster, G.R. 1993. "Congressional redistricting and African-American representation in the 1990s – An Example from Alabama." *Political Geography* 12: 549–64.

Webster, G.R. 1995. "Review of *Quiet Revolution in the South – The Impact of the Voting-Rights Act, 1965–1990* – edited by C. Davidson and B. Grofman." *Social Science Quarterly* 76: 474–75.
Webster, G.R. 1997. "The potential impact of recent Supreme Court decisions on the use of race and ethnicity in the redistricting process." *Cities* 14: 13–19.
Webster, G.R. 2000a. "Playing a game with changing rules: Geography, politics and redistricting in the 1990s." *Political Geography* 19: 141–61.
Webster, G.R. 2000b. "Women, politics, elections, and citizenship." *Journal of Geography* 99: 1–10.
Webster, G.R. 2002. "The US presidential election and the *Bush v. Gore* Supreme Court decision." *Political Geography* 21: 99–104.
Webster, G.R. 2004. "Representation, geographic districting, and social justice." *Journal of Geography* 103: 111–26.
Webster, G.R. 2007. "Social justice and voting rights in the South." *Southeastern Geographer* 47: 107–10.
Webster, G.R., T. Chapman, and J. Leib. 2010. "Sustaining the 'societal and scriptural fence': Cultural, social, and political topographies of same-sex marriage in Alabama." *Professional Geographer* 62: 211–29.
Webster, G.R. and J.I. Leib. 2001. "Whose South is it anyway? Race and the Confederate battle flag in South Carolina." *Political Geography* 20: 271–99.
Webster, G.R. and J.I. Leib. 2002. "Political culture, religion, and the Confederate battle flag debate in Alabama." *Journal of Cultural Geography* 20: 1–26.
Webster, G.R. and N. Quinton 2010. "The electoral geographies of two segregationist ('Jim Crow') referenda in Alabama." *Political Geography* 29: 370–80
West, W.J. 2005. "Regional cleavages in Turkish politics: An electoral geography of the 1999 and 2002 national elections." *Political Geography* 24: 499–523.

PART II
Electoral Geography Outside of the US

Chapter 3
Electoral Systems, Geography, and Political Behaviour: United Kingdom Examples

Ron Johnston and Charles Pattie

Geography is at the core of most systems used to elect legislators to represent territorially-defined constituencies. Defining and then operating within such constituencies poses challenges to both the political parties and the electorate not only because they structure the choices that voters make but also because their spatial configuration can strongly influence the electoral outcome – how votes are translated into seats. These characteristics are not always realized, except in the case of single-member, plurality constituency systems (often termed first-past-the-post or FPTP) where the potential for gerrymandering is great – not only intentionally, as in the USA, but also unintentionally because of the rules that non-partisan districting bodies have to apply, as Gudgin and Taylor (1979) showed in a classic contribution to electoral geography and political science. Constituency definition can also have major impacts on the operation and outcome of contests using other types of electoral system, however.

The United Kingdom offers a fascinating case study for appreciating these geographical components of electoral operation, because of the wide variety of systems deployed: indeed, examples of most of the main types of system can be found there (Electoral Reform Society, 2007c; Ministry of Justice, 2008). The basic system, used for elections to the House of Commons (the lower house of the UK Parliament), is first-past-the-post; this is also used in elections to local governments in many parts of England and Wales, though in some areas (such as the English metropolitan boroughs) each ward returns three councillors at an election, with voters having three votes. For the election of city mayors in England – including the Mayor of London – a preferential voting method termed the supplementary vote in single member constituencies is used, and a similar system, the Alternative Vote (known in North America as instant run-off), has been proposed for electing the House of Commons – if approved in a national referendum to be held in May 2011.

In addition to those using single-member constituencies, all of the three other main types of electoral systems are also used, at least in some parts of the United Kingdom. List systems, designed to produce proportional representation, are widely used in many parts of Europe, and one – the closed party list in multi-member constituencies – was adopted by the UK Parliament, in order to meet European Union requirements, for elections to the European Parliament from

1989 on, in England, Scotland and Wales, but not Northern Ireland. (Oelbermann et al., 2010, illustrate the considerable variety of systems used across the 27 member states for elections to that Parliament, consistent with the proportional representation requirement.) Another multi-member constituency system that involves preferential voting – the Single Transferrable Vote – that can, in many circumstances, produce a result broadly consistent with the principle of proportional representation, is used for elections to the Northern Ireland Assembly, to the European Parliament from Northern Ireland, and to local governments in Northern Ireland and Scotland. Finally, versions of the hybrid system often known as Multi-Member Proportional or Additional Member Proportional, in which electors have two votes, one for legislators to represent single-member constituencies (elected by FPTP) and another to elect members of a party list in multi-member constituencies, are used for elections to the Scottish Parliament, the National Assembly of Wales, and the London Assembly.

Examples of all of these systems are used here to illustrate the role of geography in their construction and operation. Having introduced their core features, we focus on the issues that they raise for both parties and voters, plus the consequences of how they translate votes into seats.

The Issues

Three main sets of issues are addressed here:

1. The geography of the system – basically how the constituencies are defined and their boundaries demarcated;
2. Issues that the parties have to address, such as how many candidates to field in some circumstances, how to spatially structure their campaigns, and what sort of advice they offer to the electors – which may be much more than 'vote for us';
3. Issues that the electors have to address, such as whether to vote tactically (Americans call this voting strategically) and what information they need to make such decisions.

One feature common to all systems is the minimum requirements principle, which is the smallest number of votes that will guarantee election in a particular constituency. This number is generally known as the quota, and it varies according to the magnitude of the constituency (i.e., the number of legislators that it elects, not the number of voters there or some other measure of its size). It is derived by the formula Q = [(total number of valid votes cast/(total number of legislators to be elected + 1)] + 1.

Thus, in a constituency returning one member in which 100,000 votes are cast, the quota is [(100,000)/(1 + 1)] + 1 = 50,001 which is, of course, a bare majority:

any candidate with that number of votes (and there can only be one) must be elected. If two legislators are to be elected, then the quota is [(100,000)/(2 +1)] + 1 = 33,334 and again any candidate with that number of votes is guaranteed election (since if there were two other candidates, only one could get more than 33,334 – if she got 33,335 votes, for instance, then the third candidate could only get a maximum of 33,331).

Basically the larger the magnitude of the constituency, the smaller the Q. It can alternatively be expressed as PCQuota = [100/(n + 1)] where n is the constituency magnitude (number of legislators to be elected). Thus in a two-member constituency, 33.3 percent of the votes will guarantee election for a candidate, in a four-member constituency 20 percent are sufficient, and in a nine-member constituency 10 percent.

Single-Member Constituency Systems

Single-member constituency systems are a minority in the world, being especially popular only in the English-speaking countries, most of them formerly part of the British Empire. Among those countries there have been moves away from that arrangement: in 1994, for example, South Africa's first fully democratic general election – and all subsequent contests – have been held using a proportional representation (PR) list system and two years later New Zealand's general election for the first time used a hybrid system, introduced after two popular referendums rejected the British system that had previously been used there (Renwick, 2010).

First-Past-The-Post (FPTP) – or Plurality – Elections

FPTP elections have been the sole means of electing members of the House of Commons only since 1951. In the nineteenth century multi-member constituencies were the norm, with electors having the same number of votes as there were seats to be filled, so that if there were two seats for a constituency the two candidates with the largest number of votes were elected. The Reform Acts of that century (1832, 1867, 1885) and of 1918 slowly replaced those seats with single-member constituencies, but it was not until 1949 that the last multi-member constituencies were removed (Rossiter et al., 1999).

FPTP elections are often promoted as straightforward, especially for the elector; the act of voting is ostensibly simple – voters choose their preferred candidate/party – and identification of the winner is similarly clear – the candidate with most votes. But the system is open to many distortions and biases, and can – in most circumstances – present the voters with difficult choices.

The results of UK general elections since 1950 have been characterized by two main features. The first is disproportionality: parties' percentages of the votes cast are not commensurate with their percentages of the legislative seats won. Some get a higher percentage of seats than votes; others get the reverse. In general, the largest party in terms of votes cast gets an even larger share of the seats and the smallest parties get lower shares of the seats than of the votes – there is in effect a 'winner's bonus', which usually means that the winning party gets a majority of seats in the House of Commons (even though no party has won a majority of votes at any election during that entire period). But this is not always the case. In 1951, for example, Labour won 48.8 percent of the votes cast (its largest share ever) and the Conservative-Unionists 48.0 percent, but the latter got 321 seats (a Parliamentary majority) to Labour's 295. At the first election held in 1974, the situation was reversed: the Conservatives won 37.9 percent of the votes compared to Labour's 37.2, but Labour got four more seats than its main opponent (but not a Parliamentary majority). And then in 2010, the Conservatives were again the largest party, with 37.4 percent of the votes, but they obtained only 47.2 percent of the seats, resulting for the first time since 1945 in the creation of a formal coalition government.

The main reason for this disproportionality is geography, specifically the geography of where each party wins its votes within the matrix of constituencies – as laid out in one of the classic works of electoral geography (Gudgin and Taylor, 1979). Take the 1951 election as the example. If Labour had won 48.8 percent of the votes in all of the 625 constituencies and the Conservatives 48.0 percent, then Labour would have won all of the seats. It didn't because too many of its votes were in the wrong places. As Gudgin and Taylor (1979) demonstrated, the key parameter in the translation of votes into seats is the frequency distribution of a party's share of the votes across all constituencies. If it wins a very large share of the votes in a small minority of the seats (65 percent in 100 of the 625, say) but a smaller than average share in a much larger set (35 percent in 350, say) then it will win all of the former but lose all of the latter. Its opponent, on the other hand, may have its votes much more efficiently distributed, winning perhaps 52 percent of the votes in 400 of the seats. The latter will then get a much better return for its share of the votes cast than the former – hence the disproportionality. And of course a small party, winning three percent of the votes, say, will get no seats at all – unless most of them are concentrated in a very small number of constituencies where it obtains a plurality (which was the case for the UK Liberal party in 1951, that gained 2.5 percent of the votes but only six MPs out of 623).

To some observers, the disproportionality inherent in the FPTP system is among its strengths – despite the occasional anomalies when the winning party in terms of votes is not also the largest in terms of seats. (Such anomalies also occurred in New Zealand at successive elections in 1978 and 1981.) The largest party is able to form a government with a workable majority in most cases but because the system tends also to magnify changes in support – a small change in percentage of the votes won is usually associated with a much larger change in share of the seats, again a function of the frequency distribution of the former (especially if it is leptokurtic: Gudgin and Taylor, 1979) – a governing party that loses popular support is thus likely to be replaced. Small parties are disadvantaged, however, but this is accepted as the price to pay for clarity in the election results. In effect, as Duverger (1954) argued in his classic study of electoral systems, FPTP is not only most appropriate for a country where two parties predominate but also ensures in most cases that this is the case.

The United Kingdom, however, no longer has a two-party system – if it ever did. It has always had some smaller parties that were strong in some parts of the country only – Nationalists and Unionists in Northern Ireland; the Scottish National Party (SNP) in Scotland; and *Plaid Cymru* (PC) in Wales – but only in Northern Ireland have these come to dominate and create a different party system. (Separate party systems in the different Provinces are common in Canada, of course.) Within Britain, the SNP and PC increased their support substantially after 1970, alongside resurgence in voter support for the Liberal party (which after a merger with the Social Democrats in 1988 became the current Liberal Democrat Party). As a result of their combined success, a multi-party system began to emerge: the Conservative and Labour combined share of the votes cast fell from over 90 percent in the 1950s to only 67 percent in 2010, but they still obtained the great majority of the seats – 86 percent in 2010. The Liberals were especially disadvantaged because their votes were too widely dispersed – at elections since 1992 they have averaged around 20 percent of the votes but less than 8 percent of the seats.

The growth of a multi-party system (with a number of other smaller parties, some of which – like the United Kingdom Independence Party (UKIP), the British National Party (BNP) and the Green Party – contest many constituencies now) means that disproportionality is even greater than in the earlier post-World War II decades. Alongside it a further feature has developed – bias. The system does not treat the two largest parties equally in the translation of votes into seats. At recent elections, Labour has benefited very substantially from this. If we take a 'fair disproportional' election system as one that translates votes into seats in the same ratio whichever party is the victor, it would be unbiased. But if one party is advantaged over the other the electoral system is biased. As a baseline, we can argue that if two parties got the same share of the votes cast, they should get the same number of seats. At the last four general elections this was not so. In 1997, if the Conservatives and Labour had been equal in vote shares, Labour would have won 82 more seats than its opponent; in 2001, it would have won 142 more; in 2005, 112; and in 2010, 54 (Johnston et al., 2006; Johnston and Pattie, 2011a).

The reasons for this bias are, in effect though not in means, the same as the classic causes of electoral abuse in the United States – malapportionment and gerrymandering (Johnston et al., 2001). Constituency boundaries in the UK are reviewed approximately every decade by non-partisan commissions operating rules laid down by Parliament. The boundaries are not put in place to advantage one party over another (although because the parties can challenge, though not determine, the boundaries during the public consultation process there are some attempts at the equivalent of gerrymandering, which are of minor importance only: Rossiter et al., 1999). But, again as demonstrated by Gudgin and Taylor (1979), because those constituencies are superimposed upon maps of support for the parties, and all constituencies are – for a variety of reasons (Borisyuk et al., 2010) – not the same size, the equivalent of gerrymandering and malapportionment emerges; and since 1992 this has been substantially to Labour's advantage (Johnston, 2002).[1]

This superimposition of the constituency map on that of voter preferences creates a geography that has become of increasing relevance to the country's political parties and voters. Of the two largest parties, the Conservatives have their greatest support among the middle classes, concentrated in the suburban and rural areas, whereas Labour's heartlands are in the inner cities and the industrial areas. For each, therefore, there are areas of the country where their support is so strong that they are almost certain to win – their 'safe seats' – and others where they are sure to lose – their 'hopeless seats'. The contest between them is thus compressed into those constituencies where neither's support base predominates – the 'marginal seats', comprising constituencies that incorporate a range of socio-economic environments. These are where the election is won and lost, and are the parts of the country where the parties and their candidates campaign most intensively. Voters in the safe and hopeless seats get relatively little attention from the parties in the months and weeks before an election whereas those in the marginals are the subject of a great deal of interest as the local party machines (increasingly carefully managed and resourced from central party headquarters) invest much time and energy into identifying their potential supporters and encouraging them to vote and, potentially, make a difference to the result. And it works; the more intensively a party campaigns in a constituency (especially a party other than one currently in government) the better the result for it there (Johnston and Pattie, 2008; Pattie and Johnston, 2009; Johnston and Pattie, 2011b).

Voters respond to this geography of campaigning, of course. Some, who feel that their vote is not needed, abstain. In particular, electors in 'hopeless seats' are markedly less likely to vote (even to use the increasingly-promoted postal voting facility introduced to increase turnout – which has fallen nationally by some 20 percentage points in recent decades) than those in the marginals. Turnout is generally lower in such constituencies, especially those that are Labour-held; potential Labour voters decide that their vote is not needed – especially if they haven't been canvassed by the party – whereas supporters of other parties decide their votes too will have no impact and so abstain. (The former is less likely to

1 In 2010 the coalition government that took office after the election in May that year introduced a Parliamentary Bill that would both reduce the number of MPs (from 650 to 600) and require the new constituencies to be drawn so that all their electorates (with two exceptions in Scotland) were within 5 percent of the national quota (some 78,000). (On predecessors to that proposal, see Johnston et al., 2009.)

be the case in Conservative areas; that party's voters are more likely to turn out whatever the local situation.)

Abstaining is only one response to the geography of the election and its campaigns. Rather than abstain in the knowledge that their vote may have no impact, some electors decide to vote for another party. Indeed, they do that in considerable numbers. The 2005 British Election Study survey, for example, found that 15 percent of all respondents who voted did not cast their ballot for their most preferred party. Of those who preferred a party other than either the Conservatives or Labour, 45 percent voted for a party other than their first choice; and of those whose first preference was neither one of those two parties nor the Liberal Democrats, fully 79 percent of them voted for other than their most preferred candidate. To some extent, this was because their preferred party did not present a candidate in the constituency, but that was not the major cause. The Conservatives, Labour and the Liberal Democrats fielded a candidate in every British constituency, for example, as did the SNP and PC in Scotland and Wales respectively. And the 'minor parties' also offered candidates in most places: of the 632 constituencies in Britain, 557 had a UKIP candidate, 337 one from the BNP, and 332 from the Greens.

The main reason for voting for a candidate other than one's first choice at a British election is what is termed tactical voting there (strategic voting in North America: for an introduction to the concept, see Cox, 1997). Many vote for their second preference because that person has a better chance of victory locally than their first choice. Thus, for example, in 1997 the two left-of-centre parties – Labour and the Liberal Democrats – wanted to end the Conservative sequence of governments that started in 1979 and so in constituencies where Labour candidates has the best chance of unseating a Conservative incumbent many Liberal Democrats voted Labour to achieve that, whereas in constituencies where the Liberal Democrats were best placed for victory the reverse was the case. Since 1997, such tactical voting has focused on preventing a Conservative return to power. In many cases, the parties themselves have promoted – implicitly if not explicitly – such tactical voting: many Liberal Democrat candidates have published leaflets with electoral data (some of it fairly spurious) claiming that only they could beat the Conservatives in their constituency (a tactic supported by some of the media, and by senior members of both the Liberal Democrat and the Labour parties – with the latter promoting what they termed 'intelligent voting'). In 2010, when all three parties believed they could win at some stage of the election campaign, as well as the promotion of tactical voting there was also 'anti-tactical-voting' campaigning: Conservatives in some areas claimed that a Liberal Democrat vote would assist the return of a Labour government, for example, whereas in other areas Labour campaigned that a Liberal Democrat vote could ensure a Conservative administration. Tactical voting is now an established feature of British elections, with some 15 percent engaged in it in 2010. As a result, instead of England having a three-party system, it has in effect three two-party systems: in about half of the constituencies, the contest is between Labour and the Conservatives, with the Liberal Democrats a poor third; and in about one-quarter each there are Conservative-Liberal Democrat contests (mainly in the non-metropolitan south of England) and, in the urban north, Labour-Liberal Democrat contests (Johnston and Pattie, 2011a)

First-past-the-post is thus a system that is open to manipulation through the cartography of its constituency map (Johnston, 2002) and through tactical voting, and election results are likely to display both disproportionality and bias. It meets what, to many, is a basic criterion for a successful electoral system – the elected legislators have a clear and individual link to a geographical segment of the relevant territory – but it fails to meet many others, such as PR. The same conclusions can be drawn about other systems that are based on single-member constituencies.

Preferential Voting in Single-member Constituencies

Whereas in FPTP electoral systems voters pick one candidate/party only to receive their support, in preferential systems they place two or more of the candidates in order of preference. The goal is to ensure that the elected candidate has majority support; FPTP in multi-party systems can result in many of the legislators being elected with a minority of the votes only (as was the case with two-thirds of the MPs elected at the 2010 general election). These preferential systems do not eliminate the disproportionality and bias characteristic of FPTP – indeed, they probably accentuate them, because of the issues involved in constituency delimitation – but they also introduce new issues for both parties and electors.

The Supplementary Vote

The supplementary vote (SV) was introduced in the late 1990s for the election of the Mayor of London and also mayors in a small number of other cities and towns where such a form of government was supported by the local population in a referendum. (It was devised by a Labour MP when his party was considering electoral reform after its third successive defeat by the Conservatives in 1987, was commended by the Plant Commission – Plant, 1993 – and subsequently recommended for the London Mayoral election: Dunleavy and Margetts, 1998; van der Kolk, Rallings and Thrasher, 2004; Rallings, Thrasher and Cowling, 2002.) Voters are presented with a ballot paper in which they indicate both their first and their second preferences. The first preferences are counted, and if one candidate has secured more than 50 percent (i.e., has achieved the quota) he/she is declared elected. If there is no outright winner, there is an instant run-off between the top two candidates. All of the others are eliminated and their second preferences allocated: if they are for one of the first two candidates, they are added to the relevant total; if they are for one of the eliminated candidates, they are discarded, and the second preferences of electors who voted for one of the run-off candidates are ignored. Of the candidates in the run-off, the one with most votes

Table 3.1 Results of the First Three Elections for the Mayor of London, Indicating the Number (and Percentage) of the First Preference Votes and the Final Number of Votes Going to the Candidates of the Two Leading Parties

	First Preferences	Final	Total Votes Cast
2000			
Labour	667,877	776,427	1,714,072
(39.0)			
Conservative	464,434	564,137	
(27.1)			
2004			
Labour	685,541	828,380	1,863,671
(35.7)			
Conservative	542,423	667,178	
(28.2)			
2008			
Labour	893,877	1,028,966	2,415,958
(37.0)			
Conservative	1,043,761	1,168,738	
(43.2)			

(first plus second preferences) is elected: this may be with a clear majority – if most of the supporters of the eliminated candidates gave their second preferences to one of the candidates in the run-off.

Table 3.1 shows the results for the three elections of the Mayor of London in 2000, 2004 and 2008. On each occasion the leading candidate failed to gain a majority of the first preference votes, but sustained his position after the allocation of second preferences. In 2000, the winning candidate (Ken Livingstone) had 69 percent of all of the votes after preferences had been allocated, but in each of the subsequent elections the victor failed to get an overall majority (Livingstone got 43 percent of the 1.9 million cast in 2004 and his opponent – Boris Johnson – 46 percent in 2008, indicating that a substantial number of second preferences in each case were wasted).

Indeed, the wasting of votes can be a major issue. In the first election for Mayor of Stoke-on-Trent, the leading two candidates – who entered the run-off stage – each got about one-fifth of the first preference votes (there were 12 candidates). The second-placed candidate was the eventual victor, but although 24,866 votes were cast for the ten eliminated candidates, only 5,964 were allocated to one of the two run-off candidates – so that the winner was elected despite 75 percent of the voters giving him neither their first nor their second preferences (van der Kolk, Rallings and Thrasher, 2004). Such extensive wasting can occur for one of three reasons: people only indicate their first preference; people mistakenly give their first and second preferences to the same party; or people give their second preferences to an eliminated candidate. (The last of these is especially likely to occur when voters are uncertain which two candidates will be in the run-off.) Data for the first two London Mayor elections allow a decomposition of the wasted votes. Thus in 2000, of the 581,761 first preferences given to candidates other than the two front-runners, 16 percent had no stated second preference, a further 7.7 percent had the second preference allocated to the same candidate as the first preference and so were eliminated, and 40.5 percent went to candidates outside the first two; thus only 35.8 percent of the potential second preferences were allocated to one of the two leading candidates. Four years later, when the same two candidates occupied the run-off positions, 42.1 percent only of the potential second preferences was allocated to one of them (van der Kolk et al., 2006). Thus whereas in each case the winner (Ken Livingstone) obtained a majority of the 'effective votes', after the allocation of second preferences he was elected by only a minority of all of those who voted (45.2 and 44.4 percent in 2000 and 2004 respectively).

This system clearly involves electors having to make more than the single decision involved under FPTP, selecting their second as well as their first preference candidate. In doing so, they have to be aware that their second preference will not count if it is given to a candidate other than one of the two who are in the run-off. How will they know who those two will be?[2] In some high profile contests opinion poll data might provide the needed information – but if it is a fairly close three-way contest that might not help. Otherwise, electors will have to rely on either their own general appreciation of the local political situation or what they are told by either the parties (in electioneering materials) or the local media – which may not be reliable. Parties and/or candidates may campaign for second preferences if they are confident of making the run-off, whereas others who are not may suggest how their first preference voters should use their second choice – which would also be the case if the supplementary vote were used to replace FPTP in Parliamentary elections.[3] In general, however, many electors may 'waste' their second preference in the absence of any definitive information/advice – which of course may be erroneous if those offering it wrongly identify the two candidates who will make the run-off.

The SV system has not been used in the United Kingdom to elect a representative body rather than a single executive, such as a mayor. If it were, it would clearly carry many of the same disadvantages as FPTP: indeed, it could well exacerbate

2 In 'true' runoff elections, where a second round of voting is held at a later date, with candidates getting less than a specified percentage of the first round votes being eliminated – as in Presidential and Legislative elections in France – this problem is absent.

3 Of course, to suggest how supporters might use their second preferences could be seen as an admission of defeat, unless it was deployed in a positive way to show how that could influence the winning candidate's policies.

them. It is designed to ensure that the victor has a majority mandate – by the allocation of second preferences if no candidate gets above the quota (50 percent of the votes cast plus one) – but by restricting the allocation of preferences only to the top two candidates on the first preference votes it prevents a relatively popular 'third party' candidate from overtaking them. Disproportionality – measured using each party's percentage of the first preference votes and of the seats – is thus likely to be as large if not larger than with FPTP, and the outcome could also be biased for the same reasons that FPTP results are. Further, because the electors are to some extent operating under a 'veil of ignorance', in that their decisions on who to support in the run-off must be made without knowing who the candidates in that run-off will be (though they may be able to estimate that – or be given an estimate by the parties and their candidates), it may well be that many of the second preferences – and hence the first preferences with which they are linked – are ineffective (or 'wasted').

The Alternative Vote

In this system (AV), rather than indicate first and second preferences only among the candidates, electors are asked to rank order them all. If a candidate gets a majority of the first preference votes, he/she is elected. If no candidate achieves the quota, however, the one with fewest first preferences is eliminated, and her/his second preferences allocated among the others; if one then exceeds the quota, the election is over. If no candidate still reaches the majority, the remaining one with fewest first preferences is eliminated and second preferences allocated – or third, if the second preference is for an eliminated candidate. The procedure continues until either one candidate reaches the quota or only two remain, in which case the one with most votes is elected. (This may not be a majority because some electors do not rank order all of the candidates.)

The alternative vote system was pioneered in Australia in 1919 as a way of combining the votes for several right-of-centre parties that otherwise might lose an FPTP election to a single left-of-centre party's candidate who failed to obtain a majority of the votes cast (McLean, 1996); it has been used since then for elections to the country's House of Representatives, as well as to several of the state legislatures (Farrell and McAllister, 2006). It places considerably more demands on the voter than even SV, especially where there is a large number of candidates (in some Australian elections, electors have been asked to rank order more than 20 candidates in a single constituency). Although voters do not need to be prescient as to which the leading two candidates on the first preferences will be, some sophisticated decision-making may be needed to ensure that their votes are not wasted, especially if their first preference is for a party very unlikely to be elected in the constituency. Tactical voting can be very effective, especially at preventing the election of a candidate with substantial, but not majority, first preference support but, unlike FPTP, the system does not discourage people from giving their first preference to a candidate who will almost certainly lose.

If the second and lower preferences are allocated, however, the order in which candidates are placed can be crucial – which is why Australian political parties distribute 'how to vote' cards to voters, suggesting a rank-ordering that is in their own candidates' best interests by reducing the chances of their main contenders.

The production of these 'how to vote' cards will almost certainly be preceded by political bargaining among the parties. If parties *A* and *B* both want party *C*'s voters' second preferences to ensure victory in as many constituencies as possible, they may well have to concede legislative concessions to party C (perhaps without divulging that they have), and failure to fulfil their promises could make it more difficult to bargain for such recommendations from party *C* to its voters at future contest. Furthermore, a sophisticated electorate could soon realize that a party that has little or no chance of victory should still get its supporters' first preference votes. Indeed, the more first preference votes that such parties obtain, the greater the probability that no party will win an outright majority on the first preferences: not only will such minor party supporters be more likely to influence the electoral outcome in a considerable number of constituencies but they are also more likely to have an impact on policy development than if they gave their first preferences to the ultimate winner (Jaensch, 1983, calls this 'second-preference government'). In addition, the more first preference votes that go to 'hopeless' parties the greater the probability that the margin between the two (or three) leading parties will be small, putting greater power into the choice of second preferences by the eliminated candidates' supporters

AV, like SV, does not produce election outcomes consistent with proportional representation. Indeed, because it favours the largest parties it is likely to generate outcomes that are even more disproportional and biased than those associated with FPTP (Johnston and Forrest, 2009a, 2009b). Nevertheless, AV was proposed as a replacement for FPTP by the UK Labour party in 2010, in an attempt to win support from the Liberal Democrat party (which favours PR using the STV system discussed below) in a possible coalition agreement after the general election then. Their argument was that because AV ensured that all MPs were elected with a majority mandate from their constituency this would enhance their legitimacy – although the *realpolitik* was that the main losers, should it be adopted, would be the Conservatives if, as was probably the case until then, Liberal Democrat candidates were more likely to get second preferences from Labour candidates in third place than were the Conservatives, and also that Labour candidates would pick up more second preference votes from third-placed Liberal Democrat candidates than the Conservatives.

Some Liberal Democrat politicians saw AV as a valuable first step on the route to PR, recognizing its advantage in eliminating some of the pressure for tactical voting; more voters would be able to indicate their 'sincere (first) preference' for the Liberal Democrats under AV without wasting their vote if their other preferences were to be taken into account. But the potential for tactical voting is present in an AV election. If, of three remaining candidates in a constituency, *B* is likely to lose to *A* after the final preferences are counted but to beat *C* – on the assumption that *B* will be in the final run-off – then *B*'s supporters should try to ensure that *C* is not eliminated before the final distribution of preferences; if enough of *B*'s supporters vote for *C* so that it enters the run-off but that *B* is not eliminated – leaving *A* and *C* in the run-off – then *B*'s chances of ultimate victory are enhanced; you want to run-off against the weaker of your two main opponents if you are sure of being at that stage yourself. Such a tactic of course depends upon detailed information being available to the voters – either from independent sources that they can employ to make their own tactical decisions or, more likely, from the parties who stand to gain from such behaviour. Alternatively, a party that knows it cannot win in a constituency may prefer to come third – so that its second preferences are then crucial in ensuring that a less-preferred candidate wins – than coming second and then being beaten.[4]

The outcome of the 2010 general election was a coalition of the Conservatives – who were strongly in favour of retaining FPTP – and the Liberal Democrats. As part of the agreement, the Conservatives agreed to support a Parliamentary Bill to hold a referendum on a change from FPTP to AV in 2011 – but reserved their right to campaign against the change when it was put to the electorate. Labour was prepared to legislate for the change without a referendum – having previously promised one if re-elected – if the Liberal Democrats entered a coalition agreement with them rather than the Conservatives, but the negotiations broke down on a number of issues and a Conservative-Liberal Democrat coalition was formed (on which see Mandelson, 2010).

Multi-Member Proportional Representation List Systems

There are two basic types of list system. In the first – known as closed-list systems – the political parties nominate a list of candidates for each constituency in rank order. Electors vote for their preferred list, and parties are then allocated a number of legislators consistent with their share of the votes. Thus in a nine-member constituency a party with 30 percent of the votes would be guaranteed three legislators (the PCQuota is 10), and the first three candidates on its list would be declared elected. The other type is known as the open-list system, in which instead of just plumping for a party the electors can, if they wish, indicate their first (and sometimes lower) preferences among their chosen party's candidates – they can only vote for one party's candidates. In these cases, if a party gets 30 percent of the votes in a nine-member constituency, the three most popular of its candidates are elected. (If an elector does not indicate her/his preferred candidate, then her/his vote is given to the top candidate on the party's list.)

---

4 If *C*'s supporters know that they cannot win in a constituency against *A*, then they might prefer that their second preferences ensure that *B* defeats *A* than take the chance that if *B* came third, its supporters would give insufficient second preferences to *C* to prevent *A*'s victory.

Table 3.2 Allocation of Seats across the Nine English Regions for the European Parliament Elections in 2009

Northeast	3
Northwest	8
Yorkshire and the Humber	6
East Midlands	5
West Midlands	6
Eastern	7
London	8
Southeast	10
Southwest	6

As open list systems are not currently used in the UK, we will not address any particular issues that they raise. Closed list systems raise three. Basically they are fairly straightforward to operate. Parties present a full field of candidates – certainly at least one more than they expect to get elected and no problems are raised if they nominate more – and they campaign for the maximum number of votes.

There are two apportionment issues, however. The first relates to the definition of constituencies. Two options are available to the system designers: a bespoke set of constituencies for the task, defined by a Boundary Commission or similar body; or using an existing set of administrative areas. The latter option was taken in the UK for elections to the European Parliament. Each of the four constituent countries was allocated a number of seats from the total – in the latest allocation, Northern Ireland was allocated three, Wales four, Scotland six and England the remaining 59. Within England, the total was then subdivided across the country's nine administrative regions, as shown in Table 3.2.

This task may seem straightforward, but there is a variety of ways in which the number of seats can be allocated – ways that have been debated for more than a century and have informed the decennial allocations of House of Representatives' seats in the United States where some important paradoxical aspects the procedures explored were generated (Balinski and Young, 2001). In the UK case, the goal was to allocate seats to regions consistent with the number of registered electors there, and the most proportional formula – the Sainte-Laguë (or Webster) was used by the UK Electoral Commission: use of another formula could have changed the allocation slightly, with the Southeast getting one more seat.[5]

This allocation immediately identifies an issue for the parties and electors. The different number of seats per region means that the quota varies spatially. In the extreme cases within England, in the Northeast region a party needs at least 25

5 An accessible, non-technical introduction to these issues is Szpiro (2010).

percent of the votes to be guaranteed one seat, whereas in the Southeast it needs only 9 percent. (Many countries have a threshold below which parties will not be allocated seats. This is usually no more than 5 percent, although in Turkey it is 10 percent; at the 2007 elections there only three parties crossed that threshold and the largest – with 46.6 percent of the votes – won 341 of the 550 seats in the legislature. On various systems, see Carey and Hix, 2011.) Smaller parties are better able to win representation in some parts of the country than others, therefore. In those where their chances are low they may decide not to present a list, or at least not to campaign hard there if they do so. Supporters of such parties, if they are aware (or made aware, presumably by other parties) of the issue may decide not to 'waste' their vote by casting it for their preferred party and instead vote for their second preference, which has a much greater chance of success. Tactical voting is sensible in some proportional representation situations, therefore.

The second allocation issue relates to the distribution of seats within each region once the votes are counted. A party will rarely get precisely the quota and thus an integer allocation of seats and so some formula has to be deployed to make the most proportional allocation. As with the allocation of seats to region, Sainte-Laguë/Webster achieves that, but in many cases, including the UK's elections to the European Parliament, the less proportional D'Hondt/Jefferson formula is used. This can create significant differences (McLean and Johnston, 2009). With D'Hondt, the total number of votes each party receives is divided by, successively, 1, 2, 3, 4 …, whereas with Sainte-Laguë the sequence is 1, 3, 5 ,7 ….[6] Seats are then allocated to the parties achieving the highest *n* ratios (where *n* is the constituency magnitude – the number of members to be elected from that constituency), as illustrated in Table 3.3 for the 2009 European elections for the Southwest England constituency (only the six largest parties are shown).

The seats are allocated sequentially according to the size of the ratios. Thus under the D'Hondt formula the first seat goes to the Conservatives, the second to UKIP, the third to the Liberal Democrats and so on. Only three parties win seats, whereas under Sainte-Laguë five do, which is a substantial difference for a constituency returning only six legislators. Indeed, across the whole of Great Britain (excluding Northern Ireland, which elects its three MEPs by a different system), the difference between the two systems in the overall outcome would have been quite substantial, as shown by Table 3.4. In effect, the D'Hondt formula favours the largest parties (the Conservatives and UKIP won 27.7 and 16.5 percent of the votes respectively, but 36.2 and 18.8 percent of the seats) whereas the smaller parties suffer (the Greens won 8.6 percent of the votes but 2.9 percent of the seats; the BNP got 6.2 and 2.9). The Sainte-Laguë formula substantially reduces that disproportionality. (For more details, see McLean and Johnston, 2009.)

6 In some Scandinavian countries 1.4 is used as the first Sainte-Laguë denominator rather than 1, which slightly disadvantages smaller parties.

Table 3.3 Allocation of Seats in the European Parliament Election of 2007 for the Southwest Region of England According to the D'Hondt and Sainte-Laguë Divisors

	Divisor			
D'Hondt	1	2	3	4
Conservative	**468,742**	**253,741**	**156,247**	117,186
Labour	118,716	59,538	39,752	29,679
Liberal Democrat	**266,253**	133,127	88,751	66,563
Green	144,179	72,090	48,060	28,544
UK Independence	**341,845**	**170,923**	113,948	78,711
British National	60,899	30,445	20,296	15,225
Sainte-Laguë	1	2	3	4
Conservative	**468,742**	**156,247**	93,748	66,963
Labour	**118,716**	39,752	23,743	16,959
Liberal Democrat	**266,253**	88,751	53,521	38,036
Green	**144,179**	48,060	28,836	20,597
UK Independence	**341,845**	113.948	68,369	48,835
British National	60,899	20,296	12,178	8,700

Note: Allocated seats are shown in bold; the figures shown in the first column are the total number of votes won by each party.

Table 3.4 The Allocation of Seats in the European Parliament Election of 2007 from Britain According to the D'Hondt and Sainte-Laguë Divisors

	D'Hondt	Sainte-Laguë
Conservative	25	21
Labour	13	13
Liberal Democrat	11	11
Green	2	7
UK Independence	13	11
British National	2	3
Scottish National	2	3
Plaid Cymru	1	2

Preferential Voting in Multi-Member Constituencies: Single Transferable Vote

One system promoted by both the Liberal Democrat party and other bodies – such as the long-established Electoral Reform Society – is the Single Transferrable Vote (STV), used for elections to the lower house of the Irish Parliament (the *Dáil Éireann*) since 1919 and to the Australian Senate since 1949. It is an extension of the AV system to multi-member constituencies. Electors rank order all of the candidates and a quota is determined (as with the above formula). Any candidate with more first preference votes than the quota is declared elected, and her/his surplus second preferences (i.e., the first preference votes additional to those needed to meet the quota) are distributed to the other candidates. When no candidate exceeds the quota, the one with fewest first preferences is eliminated, as in AV, and second (or lower) preferences distributed. The process of successive elimination is continued until sufficient candidates exceed the quota and the constituency's full complement of legislators is determined.

One feature of STV that commends it to many proselyts is that it allows voters choice of candidates within parties as well as between them (assuming that a party fields more than one!) This to some extent removes the selection of legislators from the party machines: in single-member systems and also closed list PR systems, the choice of candidates by the parties ensures that the electorate has little influence over who *can* represent them (unless the parties run open primary contests for selection of candidates prior to the election itself). Voters can thus, for example, punish a party's incumbent legislator and replace her/him by a preferred alternative representing the same party.

In some circumstances, the outcome of an STV contest approaches PR, depending on two aspects of the geography of the constituencies deployed. First, as in list systems, the greater the constituency magnitude (*n* – the number of members to be elected) the smaller the quota and the greater the electoral prospects of small parties. Secondly, whether the constituency returns an odd or even number of members can be very influential, especially in small-magnitude constituencies. For example, in a four-member constituency the quota is 20 percent, whereas with three members it is 25 percent. Where two parties predominate in an area, if one of them has less than 46 percent of the votes there, it would prefer it to have a four-member constituency, as it is likely to win two of the four seats, whereas if it has more than 46 percent it would prefer three members, of which it would gain two (Gallagher, 1974). Of course, it is not easy in most circumstances to gerrymander constituency magnitudes so as to have even-numbers where a party is relatively weak and odd-numbers where it is strong, but the potential is there and the outcome may eventuate without it being intended (a nonpartisan gerrymander) – as has been the case in *Dáil* elections (Paddison, 1976; Parker, 1986).

A further issue that parties have to address in STV systems is the number of candidates to field in each constituency as well how to campaign for their support locally. A party may be under-represented if it fields too many candidates: for example, if it expects to get 36 percent of the first preferences in a five-member constituency (quota, 16.7 percent), it would anticipate winning two seats. However, if it fielded four candidates several of them might be eliminated early because they got relatively few first preferences, and it could finish with only one being elected. In general parties should field only one candidate more than the number of seats they expect to win.

Even with a sensible selection of candidate numbers, however, a party can still fail to win a number of seats commensurate to its total number of first preferences. It is for this reason that parties seek to segment their first preference votes, by encouraging different groups of their supporters to give their first preferences to different candidates from among the slate. One common measure of such segmentation – notably in elections in Ireland, where STV has been in use for nearly a century – is geographical; voters for party *A* in different parts of a multi-member constituency are encouraged to give their first preferences to a particular candidate. Indeed, candidate selection may also be geographically-based, with *A* selecting candidates with homes in different parts of the constituency in order to promote a 'friends-and-neighbours' effect, with each candidate developing her/his own bailiwick within the territory (Parker, 1982, 1983). This practice is clearly illustrated by a leaflet used by the Democratic Unionist Party (DUP) in one six-member constituency for the 2007 Northern Ireland Assembly elections. The DUP fielded three candidates: it asked its supporters in one group of 10 polling booths to rank order them *a*, *b* and *c* (in the first three places, of course); in another 7 to use the order *c*, *b*, *a*; and in a third group of nine to put them in the order *b*, *a*, *c*. Such vote management strategies may involve asking supporters to transfer their first preference from a much-admired local candidate who would otherwise get a large personal vote to the party's other candidates. *Sinn Féin* did this successfully in the West Belfast constituency at that election. Gerry Adams, the party leader, got 6,029 of the party's 23,631 first preference vote total there, and *Sinn Féin's* other five candidates each got between 4,250 and 4,750 first preferences. Overall, *Sinn Féin* got 70 percent of the first preference votes (the PCQuota in a six-member constituency is 14.3) but a careful strategy ensured that all five candidates were elected, when only four were guaranteed. Similar strategies were successful elsewhere, with the DUP winning four of the six seats in another constituency with only 50 percent of the first preferences – in part because too many of a *Sinn Féin* candidate's second preferences were non-transferable (i.e., his supporters did not give a full list of preferences), preventing a candidate of the other nationalist party winning the sixth seat when preferences were allocated (Electoral Reform Society, 2007a).

That STV produces outcomes that are often consistent with PR also means that it is unlikely in most cases that a single party will have a majority in the relevant legislature. This is illustrated by the case of Scottish local government, where the method of election was changed from FPTP between 2003 and 2007. In the last FPTP elections, a single party obtained a majority in 20 of the 32 local authorities, but in the first STV election this outcome occurred in just five (the majority 'party' in three was others/independents; Labour won a majority in Glasgow, with 43.3 percent of the first preference votes, and in North Lanarkshire, with 49.3 percent: Electoral Reform Society, 2007b). Overall, in 2007 STV ensured that 74 percent of voters saw their first choice candidate elected, compared to 52 percent who voted for the winning candidate in 2003 under FPTP.

The Hybrids: MMP/AMP and their Variants

A hybrid electoral system, combining first-past-the-post in single member constituencies with party list PR in multi-member constituencies, was introduced in (then West) Germany in 1949 to produce outcomes that included both representation from small, identifiable territories with a top-up system that moves the overall result towards PR. It was adopted in New Zealand in 1993 in its 'purest' form and has since been adopted in a number of other places in a variety of forms (Shugart and Wattenberg, 2001; Colomer, 2004; Lundell, 2009).

At the most recent New Zealand general election in 2008, the country was divided into 70 single-member constituencies where FPTP was used. A further 52 members of its House of Representatives were elected from a single (nation-wide) list constituency. Each party's share of the total number of seats in the House was determined by the latter contest, so that a party with 31 percent of the votes would be allocated 39 seats. If it won 28 seats in the FPTP contests, they would count towards the 39 and a further 11 would then be allocated from the list presented by the party in the other component of the contest. (If it won more seats in the FPTP contests than the list vote allocation entitled it to, the party would retain those extra seats and the size of the House of Representatives would be expanded until the next election. According to this practice two extra seats were allocated after the 2008 election, the Maori party getting five seats rather than three.) The result of MMP elections is seen by some as the 'best of both worlds' – an overall PR outcome but with a substantial number of legislators representing single-member constituencies.[7]

The London Assembly

This system was adopted in 1999 for election of the London Assembly, which has 25 members: 14 are elected by FPTP from single-member constituencies and the remaining 11 from a London-wide closed list PR contest. There is no *überhangmandat*, however, so the overall pattern of representation in the Assembly may deviate somewhat from PR. That was indeed the case after the 2008 elections, when with 37.4 and 28.0 percent of the FPTP votes respectively the Conservatives and Labour won all 14 single-member constituency seats (8 and 6 respectively). In the list contest, those two parties won 34.5 and 27.1 percent of

7 There is one major quirk to the system. If a party wins more first-past-the-post seats than its proportional entitlement, as in Germany it retains them and the relevant number of MPs is added to the overall Parliament, which increases in size until the next election: this is known as the *überhangmandat*. In New Zealand, parties only qualify for seats in the House if they either win at least 5 percent of the list votes or win at least one first-past-the-post seat, which militates against the smaller parties gaining representation and so fragmenting the Parliament – which might happen under 'pure' PR (as in Israel).

the votes. Parties getting less than 5 percent of the votes were not entitled to seats and the Conservatives and Labour were allocated a further three and two seats respectively, with the remaining six going to: the Liberal Democrats (11.2 percent of the votes, three seats); the Greens (8.3 and 2); and the BNP (5.3 and 1) The overall outcome was thus close to proportionality.

That London Assembly election illustrates the issues that parties have to face under such an electoral system – especially the smaller parties. Should they contest the single-member constituencies (which in London are very large, averaging 360,000 voters), knowing they are unlikely to win any of those seats, or should they focus their attention on the list contest? If they choose the latter option, then do they advise their supporters how to vote in the FPTP contests, where most of their potential voters will presumably have to vote tactically for their second-preference given that their first is not available? If they don't, then the electors will have to make the decision without any such advice or information: tactical voting may be forced upon them – and may be especially difficult if their second preference party is also unlikely to win in their single-member constituency (a situation that could face a Green supporter whose second preference is the Liberal Democrats).[8]

The National Assembly of Wales

As part of an agreement between several of the political parties – but not the Conservatives – prior to the enactment of a Scottish Parliament in 1998, a form of MMP/AMP was accepted as the electoral system to be deployed there – and it was then also adopted for elections to the National Assembly of Wales. Because some of the parties involved in those discussions were committed to PR as the basis for elections, the Labour party – which expected to be the largest in each of the new institutions, and some of whose leading members involved in the negotiations were not averse to PR – agreed to the new system, believing that if it did not gain a majority in the new legislatures it would at least be the largest party and able to dominate a coalition government of the centre-left.[9]

8 The Liberal Democrats, the Greens and UKIP contested all 14 constituencies in 2008, so that their supporters had to decide whether to vote for them in the almost certain knowledge it would be a 'wasted' vote or to vote tactically for their second choice. The BNP contested only one constituency, so its supporters had to vote tactically in the other 13.

9 Labour formed a coalition administration with the Liberal Democrats after the first election to the National Assembly of Wales in 1999; it was able to form a single-party administration after the 2003 elections, having just 30 of the 60 seats, but it gained only 26 seats at the 2007 elections and entered a coalition with *Plaid Cymru*. In Scotland, Labour was the largest party in coalitions with the Liberal Democrats after the 1999 and 2003 elections, but in 2007 the Scottish National Party was the largest party by one seat, and formed a minority administration.

The system adopted varied from that used in New Zealand and London, however. Instead of having a single PR list election that would determine the overall composition of the legislature, Scotland and Wales were divided into a number of regions, each of which had a separate list election. Thus in Wales, for example, the 40 constituencies employed for election to the House of Commons were used to elect the FPTP members of the National Assembly (AMs). These were grouped into four regions, each of which elected four further members from a list contest. The list votes suggested the overall party composition of the region's complement of AMs. Thus if a party won 33 percent of the list votes in a region that was to return 12 AMs, it would be entitled to 4 members. If it had five of its candidates elected from the FPTP contests, however, there was no *überghangmandat*, and so it was over-represented, with one of its opponents consequently under-represented.

This operation of this system is illustrated by the results of the 2007 Assembly election, for each of the five regions (Table 3.5). The Mid and West Wales region returned 12 AMs, so the PCQuota was 7.7. Four parties qualified for seats: *Plaid Cymru* (which won 31 percent of all List votes, but 36.2 percent of the total won by the four – 85.6 percent – that had quota places) was entitled to five; the Conservatives three; and Labour and the Liberal Democrats two each. *Plaid Cymru* had won four of the eight FPTP seats, and so was allocated one further from its list of candidates; the Conservatives won two FPTP contests, and so were also allocated a list seat; the Liberal Democrats won in two constituencies and so gained no list AMs; with no FPTP victories, Labour got two list seats.

Whereas the result in that region was consistent with the role of the top-up region element (as the list component is generally referred to) allocating seats to parties that perform well in the list contest but win no FPTP seats, that was not the case in the South Wales West region. There Labour won 35.8 percent of the list votes, which meant a theoretical entitlement to four seats (the quota in that 11-seat region was 8.3, and the four parties with quotas together obtained 82.0 percent of the list votes). But it won all seven FPTP seats, and so was over-represented by three AMs, meaning that the other three parties together obtained four of the seats rather than seven. Disproportionality was a consequence of the distribution of votes and the way that these were translated into seats by the system's geography.

This example shows not only that the Scottish and Welsh version of MMP/AMP does not necessarily produce an outcome consistent with PR – which the New Zealand version does, because of the much larger, national list component – but also that it can create both a situation of voter uncertainty and the potential for tactical voting. In New Zealand electors can vote for their preferred party of government in the list contest, knowing that because the composition of the House of Representatives will reflect the distribution of list votes, then no vote will be wasted (unless their preferred party is not going to meet the threshold of 5 percent of the votes to be entitled to Parliamentary representation); they can then give their FPTP vote to the candidate they think best able to represent the constituency,

Table 3.5 Seats and Votes by Region at the 2007 Elections to the National Assembly of Wales

Party	C	PC	LD	L	BNP	UKIP	G	∑
North Wales								
Const. Seats	1	3	0	5	0	0	0	9
List Votes (%)	25.6	25.7	7.8	26.4	5.1	4.1	2.9	
List Seats	2	1	1	0	0	0	0	4
Total Seats	*3*	*4*	*1*	*5*	*0*	*0*	*0*	*13*
Mid and West Wales								
Const. Seats	2	4	2	0	0	0	0	8
List Votes (%)	22.9	31.0	13.3	18.4	2.9	3.8	4.0	
List Seats	1	1	0	2	0	0	0	4
Total Seats	*3*	*5*	*2*	*2*	*0*	*0*	*0*	*12*
South Wales East								
Const. Seats*	1	0	0	6	0	0	0	8*
List Votes (%)	20.0	13.6	11.0	35.8	4.7	4.6	2.8	
List Seats	1	2	1	0	0	0	0	4
Total Seats	*2*	*2*	*1*	*6*	*0*	*0*	*0*	*12**
South Wales Central								
Const. Seats	1	0	1	6	0	0	0	8
List Votes (%)	21.7	15.5	14.0	34.0	3.8	3.7	3.8	
List Seats	2	2	0	0	0	0	0	4
Total Seats	*3*	*2*	*1*	*6*	*0*	*0*	*0*	*12*
South Wales West								
Const. Seats	0	0	0	7	0	0	0	7
List Votes (%)	16.1	17.7	12.4	35.8	5.5	3.6	3.8	
List Seats	1	2	1	0	0	0	0	4
Total Seats	*1*	*2*	*1*	*7*	*0*	*0*	*0*	*11*
TOTAL								
Const. Seats*	5	7	3	24	0	0	0	40*
List Votes (%)	21.5	21.0	11.7	29.6	4.3	4.0	3.5	
List Seats	7	8	3	2	0	0	0	20
Total Seats	*12*	*15*	*6*	*26*	*0*	*0*	*0*	*60**

Note: * an Independent candidate won one of the constituency seats. Key: Const. Seats – the number of constituency seats won by each party; C – Conservative; PC – *Plaid Cymru*; LD – Liberal Democrat; L – Labour; BNP – British National Party; UKIP – United Kingdom Independence Party; G – Green; ∑ – Total.

knowing that whoever wins there will not alter the overall composition of the legislature. They can vote sincerely in the list contest.

That is not the case in the British version, as illustrated by South Wales West. There all of the list votes cast for Labour are wasted, because it can get no seats additional to the seven won in the FPTP contests, unless it gets some 80 percent of the list votes. As that is extremely unlikely, in order not to waste their list votes, those who would otherwise have supported Labour should vote for their second preference party – the one they would most like to see in the Assembly alongside Labour (which would probably boost the representation of either or both of the Liberal Democrats and *Plaid Cymru* rather than the Conservatives.). Of course, they should only do that if they know that Labour is going to win all – or most – of the FPTP contests, and they cannot be certain of that even if the other parties canvassing for their second preference list votes suggested that they will. The rational thing to do in such a circumstance is to vote a 'split ticket' – voting for their first preference (Labour) in the FPTP contest but their second in the List. (On split ticket voting at those contests, see Johnston and Pattie, 2002, and Johns et al. 2010.) But such rational behaviour depends on perfect information and a sophisticated understanding of how the system works in such circumstances: less than a third had the latter (none could be absolutely certain of the former) and so voted a split-ticket, with the remainder in effect wasting their list vote. Supporters of the other parties, on the other hand, should have voted sincerely in the list contest and tactically in the FPTP contests – supporting the candidate most likely to defeat Labour (again calling for information and sophistication).

Although voters in the three South Wales regions might have been fairly sure that Labour was going to win virtually all of the FPTP contests there, and so they should structure their choices in the two contests accordingly, the likely outcome at the regional level may have been less clear-cut in the other regions – so different strategies might have been called for there. In this modified version of MMP/AMP, it is very difficult for the voters to make rational decisions that will maximise the efficacy of their two votes.

Alternative Voting (AV)

A further version of MMP/AMP has been debated in the UK over the last 15 years. In its 1997 general election manifesto, the Labour party promised to hold a referendum on the electoral system – part of its uncertainty as to whether it would win an outright victory and so in preparation for a potential coalition deal with the Liberal Democrats. After the election (which it won with a very large majority) it established an independent commission to recommend an alternative system that met certain criteria ('broad proportionality; the need for stable government; an extension of voter choice; and the maintenance of a link between MPs and geographical constituencies' – which the Commission recognised in its report were 'not entirely compatible'). The Commission recommended a version of MMP/AMP, but no referendum was ever held (Jenkins, 1998).

The system was similar to that adopted for Scotland and Wales with one major difference – the 'top-up' regions were even smaller. The country was to be divided into 80 regions (eight in Scotland, five in Wales, two in Northern Ireland, and the remaining 65 in England): each would comprise a number of constituencies returning MPs via FPTP and would also return no more than one or two further MPs based on a list vote. This small list component meant that the system would almost certainly not return a House of Commons whose composition was consistent with PR, and the large number of FPTP constituencies would almost certainly see the continuation of the disproportionality and bias typical of the system it was supposed to replace. The list votes would provide some compensation to the parties that came second or perhaps third in most regions but gained no (or very few) FPTP seats there – and voters would be placed in the same tactical voting uncertainty as those in some Welsh regions as discussed above.

The system has recently been revived, however, in a slightly revised – and even more complex – form; some senior members of the Labour party suggested its adoption, with the addition that AV would be used rather than FPTP in the constituency contests, on the grounds that so doing would ensure that the constituency MPs would have a majority mandate while the list component would be a slight concession to PR supporters (especially the Liberal Democrats, with whom Labour might enter a coalition). The elector would be faced with an even more sophisticated set of choices to make – in which, in the absence of the needed information, rational behaviour would be very difficult. On the assumption that their preferred party would get their first rank in the AV contest, they would then have to decide tactically which party should get their second preference – and then whether their preferred, their second preference, or another party still should get their list vote. In certain circumstances giving their three votes to three different parties may be the rational decision – but whether that could be reached (even by a party advising its first preference supporters) before the event is very unlikely.

Conclusions

This chapter has not been structured to present an evaluation of the merits and disadvantages of various electoral systems, let alone suggest that any one is superior to the others. As has been stressed by many authors, there is no perfect system, none that performs satisfactorily against the major criteria against which they can be assessed (see, for example, Hix et al., 2010). Choosing an electoral system thus involves a careful selection of the one that best meets the needs of the particular case.

One feature that virtually every electoral system shares is that its geography is open to manipulation in one or more ways – in the drawing of constituency boundaries, for example, and in the practice of tactical voting (which may be discouraged by the patterning of constituencies). In this chapter we have focused on that aspect of various electoral systems currently either being deployed or proposed for deployment within the United Kingdom. In so doing we have emphasized various aspects of the crucial role that geography plays in the creation, operation and electoral consequences of those electoral systems.

References

Balinski, M.L. and H.P. Young. 2001. *Fair Representation: Meeting the Ideal of One Man, One Vote*. Washington, DC: The Brookings Institution.

Borisyuk, G., C. Rallings, R.J. Johnston, and M. Thrasher. 2010. Parliamentary constituency boundary reviews and electoral bias: How important are variations in constituency size? *Parliamentary Affairs*, 63: 4–21.

Carey, J. and S. Hix. 2011. The electoral sweet spot: Low magnitude proportional electoral systems. *American Journal of Political Science*, 55: 383–97.

Colomer, J.M. (ed.) 2004. *Handbook of Electoral System Choice*. Basingstoke: Palgrave Macmillan.

Cox, G.W. 1997. *Making Votes Count: Strategic Coordination in the World's Electoral Systems*. Cambridge: Cambridge University Press.

Dunleavy, P. and H. Margetts. 1998. *Report to the Government Office for London: Electing the London Mayor and the London Assembly*. London: London School of Economics, Public Policy Group.

Duverger, M. 1954. *Political Parties*. London: Methuen.

Electoral Reform Society. 2007a. *STV in practice: A Briefing on the Northern Ireland Assembly Election 21 March 2007*. London: Electoral Reform Society.

Electoral Reform Society. 2007b. *Local Authority Elections in Scotland*. London: Electoral Reform Society.

Electoral Reform Society. 2007c. *Britain's Experience of Electoral Systems*. London: Electoral Reform Society.

Farrell, D.M. and I. McAllister. 2006. *The Australian Electoral System: Origins, Variations and Consequences*. Sydney: University of New South Wales Press.

Gallagher, M. 1974. Disproportionality in a proportional representation system. *Political Studies*, 23: 501–13.

Gudgin, G. and P.J. Taylor. 1979. *Seats, Votes and the Spatial Organisation of Elections*. London: Pion.

Hix, S., R.J. Johnston, and I. McLean. 2010. *Choosing an Electoral System*. London: The British Academy.

Jaensch, D.M. 1983. *The Australian Party System*. Sydney: Allen & Unwin.

Jenkins, Lord. 1998. *Report of the Independent Commission on the Electoral System*. London: HMSO, CM-4090I.

Johnston, R.J. 2002. Manipulating maps and winning elections: Measuring the impact of malapportionment and gerrymandering. *Political Geography*, 21: 1–32.

Johnston, R.J. and J. Forrest. 2009a. Electoral disproportionality and bias under the alternative vote: Elections to Australia's House of Representatives. *Australian Journal of Political Science*, 44: 521–28.

Johnston, R.J. and J. Forrest. 2009b. Geography and election results: Disproportionality and bias at the 1993–2004 elections to the Australian House of Representatives. *Geographical Research*, 47: 95–108.

Johnston, R.J., I. McLean., C.J. Pattie, and D.J. Rossiter. 2009. Can the boundary commissions help the Conservative party? Constituency size and electoral bias in the United Kingdom. *The Political Quarterly*, 80: 479–94.

Johnston, R.J. and C.J. Pattie. 2002. Campaigning and split-ticket voting in new electoral systems: The first MMP elections in New Zealand, Scotland and Wales. *Electoral Studies*, 21: 583–600.

Johnston, R.J. and C.J. Pattie. 2008. How much does a vote cost? Incumbency and the impact of campaign spending at English general elections. *Journal of Elections, Public Opinion and Parties*, 18: 129–52.

Johnston, R.J. and C.J. Pattie. 2011a. The British general election of 2010: A three-party contest or three two-party contests? *Geographical Journal*, 177: 17–26.

Johnston, R.J. and C.J. Patti. 2011b. The local campaigns and the outcome. In N. Allen and J. Bartle, eds, *Britain at the Polls, 2010*. London: Sage Publications.

Johnston, R.J., C.J. Pattie, D. Dorling, and D.J. Rossiter. 2001. *From Votes to Seats: The Operation of the UK Electoral System since 1945*. Manchester: Manchester University Press.

Johnston, R.J., D.J. Rossiter, and C.J. Pattie. 2006. Disproportionality and bias in the results of the 2005 general election in Great Britain: Evaluating the electoral system's impact. *Journal of Elections, Public Opinion and Parties*, 16: 37–54.

Johns, R., D. Denver, J.K. Mitchell, and C.J. Pattie. 2010. *Voting for a Scottish Government: The Scottish Parliament Election of 2007*. Manchester: Manchester University Press.

Lundell, K. 2009. *The Origin of Electoral Systems in the Postwar Era: A Worldwide Approach*. London: Routledge.

McLean, I. 1996. E.J. Nanson, social choice and electoral reform. *Australian Journal of Political Science*, 31: 369–85.

McLean, I. and R.J. Johnston. 2009. When is proportional not proportional? Great Britain's 2009 elections to the European Parliament. *Representation*, 45: 349–55.

Mandelson, P. 2010. *The Third Man*. London: Harper Collins.

Ministry of Justice, 2008. *The Governance of Britain – Review of Voting Systems: The Experience of New Voting Systems in the United Kingdom since 1997*. London: HMSO, CM-7304.

Oelbermann, K.-F., A. Palomares and F. Pukelsheim. 2010. The 2009 European Parliament elections: From votes to seats in 27 ways. *European Electoral Studies* 5: 148–82.

Paddison, R. 1976. Spatial bias and redistricting in proportional representation electoral systems: A case study of the Republic of Ireland. *Tijdschrift voor Economische en Sociale Geografie*, 67: 230–40.

Parker, A.J. 1982. The 'friends and neighbours' effect in the Galway West constituency. *Political Geography Quarterly*, 1: 243–62.

Parker, A.J. 1983. Localism and bailiwicks: The Galway West constituency in the 1977 general election. *Proceedings of the Royal Irish Academy, Section C*, 83: 17–36.

Parker, A.J. 1986. Geography and the Irish electoral system. *Irish Geography*, 19: 1–14.

Pattie, C.J. and R.J. Johnston. 2009. Still talking, but is anybody listening? The changing face of constituency campaigning in Britain, 1997–2005. *Party Politics*, 15: 411–34.

Plant, R. 1993. *Report of the Working Party on the Electoral System*. London: The Labour Party.

Rallings, C., M. Thrasher and D. Cowling. 2002. Mayoral referendums and elections. *Local Government Studies*, 28 (4): 67–90.

Renwick, A. 2010. *The Politics of Electoral Reform*. Cambridge: Cambridge University Press.

Rossiter, D.J., R.J. Johnston and C.J. Pattie. 1999. *The Boundary Commissions: Redrawing the UK's Map of Parliamentary Constituencies*. Manchester: Manchester University Press.

Shugart, M.S. and M. Wattenberg (eds) 2001. *Mixed-Member Electoral Systems: The Best of Both Worlds?* Oxford: Oxford University Press.

Szpiro, G. 2010. *Numbers Rule: The Vexing Mathematics of Democracy from Plato to the Present*. Princeton, NJ: Princeton University Press.

van der Kolk, H., C. Rallings and M. Thrasher. 2004. Electing mayors: A comparison of different electoral procedures. *Local Government Studies*, 30: 589–608.

van der Kolk, H., C. Rallings and M. Thrasher. 2006. The effective use of the supplementary vote in Mayoral elections: London 2000 and 2004. *Representation*, 42: 91–102.

Chapter 4
Spatial Regression for Electoral Studies: The Case of the Italian *Lega Nord*

Michael Shin and John Agnew

Spatial regression, and more generally, spatial analysis, are premised upon the fact that spatially referenced data are "special" (O'Sullivan and Unwin 2003). Spatially or geographically referenced data refer to data that are collected on the basis of identifiable locations or places. What makes spatial data special is that they tend to be biased by the spatial interdependence (or linkages) of places for which data are collected, and frequently, because the values of individual observations are not independent, but are correlated with each other. Such biases are usually manifest as clusters or concentrations of similar values at the local scale, which in turn can lead to notable regional differences and variations at a wider scale of analysis. Detecting and evaluating such geographic biases, referred to as *spatial dependence* or *spatial autocorrelation* at the local level, and *spatial heterogeneity* at the regional scale, are the primary foci of contemporary spatial analysis.

This chapter outlines and discusses how spatial analysis and spatial regression techniques can inform and extend ecological (i.e., aggregate data) analyses of voting behavior. The first section provides an overview of the key methodological issues surrounding the evaluation of spatial autocorrelation and spatial heterogeneity in geographic data sets. Next, a worked example that focuses on the recent success of the *Lega Nord* is presented. This party, claiming to represent the interests and identities of northern Italians in a polity that the party's leaders see as biased in favor of the Italian South, has achieved considerable success in recent Italian national and local elections. In coalition with the main national center-right party, the PdL, the *Lega* has formed part of Italian national governments in 2001–6 and since 2008. In recent years the *Lega* has become obsessed with seeing Italy adopt some sort of fiscal federalism, with local governments becoming much more important than the national one in raising revenues and making expenditures across a range of functions, even though in the late 1990s it had been avowedly separatist. It balances precariously between being a party of government, a protest movement akin to the Tea Party movement in the US, and a regional party similar to those in Catalonia and Scotland. The chapter concludes with a summary discussion of various issues that arise when using the spatial analytic and geo-statistical techniques to examine the geography of support for the *Lega Nord*.

Geographic Information Systems, Spatial Analysis, and Spatial Regression

The influences of geography and geographic context upon both natural and social processes and phenomena, ranging from precipitation to voting behavior, are topics of increasing interest. Thanks to recent advances and innovations in information technology (e.g., Google Earth), levels of geographical awareness and curiosity have increased dramatically in recent years. One special class of information technology that has been central to this shift towards geographic thinking is geographic information systems (GIS). Briefly, GIS is a special type of software that integrates data from various sources, and permits such data and the relationships between variables to be visualized as maps. GIS also serves as a platform for more robust analytical techniques, such as exploratory spatial data analysis (ESDA) and spatial regression analysis.

The key requirement for using GIS for spatial analysis is access to geo-referenced data. The terms 'geocoded', 'spatially referenced' and 'geospatially referenced' are interchangeable with the term 'geo-referenced'. Geo-referenced data refer to data that can be associated with identifiable locations (e.g., geographic coordinates, addresses, postal codes, regions, *comuni*). As noted above, the locations where data are collected often introduce, shape or produce place-specific biases. Moreover, data that are collected from proximate locations often share certain traits, for instance, similarity in type, size or magnitude. In statistical terms, the value of a variable at one location is often correlated with nearby values, and is not independent as classical probability theory necessitates.

When geo-referenced data are mapped, such biases are usually manifested as spatial clusters or 'hotspots' of similar values at the local scale, and can also result in notable regional-level variations. Detecting and evaluating such geographic biases, referred to formally as *spatial dependence* or *spatial autocorrelation* at the local level, and *spatial heterogeneity* or *spatial variation* at the regional scale, are the objectives of spatial analysis. Moreover, the results from such analyses serve to inform and refine subsequent analyses, models and avenues of inquiry.

For example, consider the series of quartile maps of support obtained in the 2008 national election for the *Partito Democratico* (PD), *Popolo della Libertà* (PdL) and *Lega Nord* (LN) at the provincial level (Figure 4.1). The PD is the main center-left party, the PdL is the main center-right party, and the LN is the northern regional party. The PD (and its political ancestors) is strong in central Italy, the PdL in the metropolitan north and the South, and the LN, not surprisingly, in the North. Levels of both high and low support for each respective party tend to be concentrated across certain areas of the Italian peninsula. Put another way, the geographic distribution of party support for the PD, PdL and LN is not random. What is more interesting, however, is how localized and/or regionalized these patterns truly are. This cannot be determined from simply "eye balling" or staring at the maps.

To test whether this visual clustering of support for each party is statistically significant, the local Moran's I statistic is calculated. Moran's I provides a formal

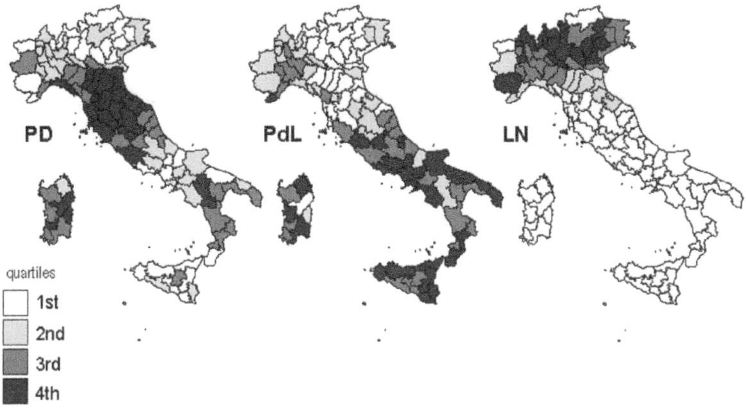

Figure 4.1 Support Obtained in the 2008 National Election for the Partito Democratico (PD), Popolo della Libertà (PdL) and *Lega Nord* (LN) at the Provincial Level

Source: Authors.

method for identifying and evaluating spatial autocorrelation (Anselin 1995a). The localized Moran's I is formally expressed as:

$$I_i = \frac{z_i}{\sum z_i^2} \sum_j w_{ij} z_j$$

where z are standardized levels of support for a given party, and w is an element in a row-standardized spatial weights matrix, W, that summarizes the spatial relationships between the Italian provinces (see Anselin 1995a, 1995b; Shin and Agnew 2008). The component, $\Sigma_j w_{ij} z_j$, is called the *spatial lag of z* and is in effect the weighted average of values that surround any given observation. In this example, W summarizes the relationship between each province and its immediate (i.e., contiguous) neighbors. Used in many statistical indicators of spatial association, spatially lagged variables provide a simple way to capture and to summarize geographic relationships in quantitative data. Inference for the local Moran's I is based upon a conditional randomization method (see Anselin 1995a; Anselin 1995b).

Note that the mean of all local Moran's I values is equal to Moran's I, the statistic most commonly used to detect the presence of spatial autocorrelation across a data set. Values of Moran's I fall between negative one and positive one, with a positive value indicating the presence of clustering and a negative value indicating a chessboard-like pattern of geographic difference (for details, see O'Sullivan and Unwin 2003). The significance of Moran's I, reported as a z-score, is based upon the assumption that values are dispersed randomly across the study area (Cliff and Ord 1981).

The values of Moran's I for the PD ($I = 0.67$), PdL ($I = 0.53$) and LN ($I = 0.92$) in the 2008 Italian general election are all high and statistically significant. In other

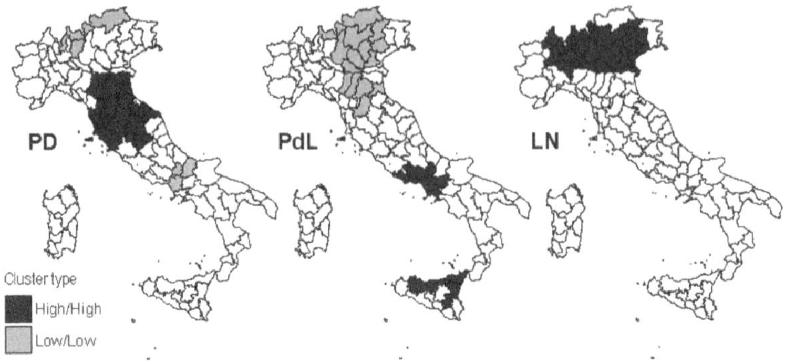

Figure 4.2 "Hotspot Maps" Showing Local Morgan's I Score for Each Party in the 2008 Italian National Election

Note: No spatial outliers were detected.
Source: Author.

words, voter support for a given party in one province is indeed associated to voter support in neighboring provinces. Since a local Moran's I score is calculated for each province, significant I values can be mapped to identify and locate "hotspots" of positive spatial autocorrelation (i.e., high values surrounded by similarly high values, or low values surrounded by similarly low values), or spatial outliers (i.e., high values surrounded by low values, or low values surrounded by high values). "Hotspot maps" for each party are presented in Figure 4.2.

Statistically significant clusters of high support for the PD, PdL and LN are shaded in dark, and clusters of low support are shaded more lightly. No spatial outliers of low or high support surrounded by dissimilarly high or low support were detected. The ability to identify the location of such hotspots and outliers provides important insights about which places may merit further exploration and explanation. Though the territorial dimensions to Italian politics are well recognized (e.g., Galli and Prandi 1970; Cartocci 1990; Shin and Agnew 2008), the spatial analytic techniques presented above serve to complement and extend such explanations of the political geography of Italy. Such methods can also be used to inform strategies for pre- and post-election surveys and interviews.

The detection of spatial autocorrelation in a data set also has implications for other statistical techniques, and in particular regression analysis. For instance, both the independent and dependent variables used in a regression may exhibit varying levels of spatial dependence. Furthermore, it is relatively common for the residuals from a regression that uses spatially referenced data to be contaminated by spatial autocorrelation, thus violating the assumption of the independence of errors. Mapping the residuals from a regression, and checking for clusters of similar values and directionality (i.e., +/-), is a relatively simple and straightforward method to check for the presence of spatial autocorrelation.

The presence of spatial autocorrelation within a data set has several implications for the parameter estimates from an ordinary least squares (OLS) regression (i.e., $y = \alpha + \beta X + \varepsilon$). When the value of a dependent variable in a given location, y_i, is influenced by or associated with spatially proximate values for the dependent variable (i.e., y_j) or independent variables (e.g., x_j), parameter estimates will be biased and inefficient (i.e., parameter estimates for the unknown true value, β, are either too high or too low). When this occurs, it is called *spatial structure* in the data. If the error term from a regression is contaminated with spatial autocorrelation, parameter estimates will be inefficient (i.e., inflated standard errors for parameter estimates). Subsequently, this can lead to incorrect inferences because possibly significant parameter estimates will not be evaluated as such. In light of recent interest and attention surrounding spatial analytic methods, formal specification diagnostics that test for the presence of spatial autocorrelation, for example, based on Moran's I, have been developed and are used widely.

To combat spatial structure in a data set, as well as the contamination of the error term with spatial autocorrelation, two spatial econometric strategies can be employed (see Anselin 1988). In the latter case, a spatial error regression model can be estimated where the spatial lag of the error term (i.e., $W\varepsilon$) is included on the right hand side of each regression (i.e., $y = \alpha + \beta X + W\varepsilon + \mu$). The spatial lag of the error term, $W\varepsilon$, is obtained by multiplying a spatial weights matrix, **W**, and the residuals, ε. The spatial lag of the error returns the weighted average of geographically proximate residual values for each observation. The inclusion of the spatial lag, $W\varepsilon$, controls for spatial autocorrelation in the error term, ε, and returns a residual term, u, that is distributed normal and independent. More importantly, the spatial error model yields more efficient parameter estimates and more reliable significance tests.

To mitigate the effects of spatial structure in a regression data set, a *spatial autoregressive* or *spatial lag* model can be estimated, in which the spatially lagged dependent variable, Wy, is included on the right hand side of the equation (i.e., $y = Wy+\alpha+\beta X +\mu$). As with the spatial error model, the spatially lagged dependent variable, Wy, is in essence the weighted average of values (e.g., vote share) from surrounding observations (e.g., provinces) and uses the same spatial weights matrix, W. Maximum likelihood estimation is used to fit such models and returns unbiased parameter estimates.

One of the assumptions underlying the spatial econometric techniques described above is that parameter estimates are homogenous, or that the relationships between variables are uniform and consistent across the data set. Depending upon the geographic context and spatial resolution of the data used in the analysis, this assumption may be untenable. For instance, it is well-established that Italian politics, economics and society exhibits a significant amount of geographic variation. Such variation, formally termed *spatial heterogeneity*, is arguably a function of how various processes operate differently in different places. In the context of regression analysis, several methods have been developed to account or to control for spatial heterogeneity, such as the inclusion of regional dummy variables, trend-surface analysis, and the estimation of separate models for subsets of a data set.

One innovative technique used to model processes that vary spatially is *geographically weighted regression (GWR)* (see Brundson et al. 1996; Fotheringham et al. 2002). Generally, GWR is concerned with estimating parameters anywhere across a geo-referenced data set. Premised upon the argument that spatially proximate observations are more likely to be similar with each other than with more distant observations, the parameter estimation in GWR gives precedence to proximate observations over more distant ones. The GWR model can be expressed as:

$$y_i = \beta_{0i}(\mathbf{u}) + \beta_{1i}(\mathbf{u})\, x_{1i} + \beta_{2i}(\mathbf{u})\, x_{2i} + \ldots + \beta_{mi}(\mathbf{u})\, x_{mi}$$

where $\beta_{0i}(\mathbf{u})$ is the parameter estimate around location **u** and is specific to that particular location. Parameter estimates can be obtained for each and every location for which there are values for the dependent (i.e., y) and independent (i.e., x) variables.

In order to obtain local parameter estimates for each and every observation in a data set, a geographical weighting scheme that describes and summarizes the spatial relations between observations must be specified. Generally, GWR uses a distance-decay function, with proximate observations having more local influence than more distant observations. Within the context of GWR, this involves specifying a *bandwidth* and a *kernel*. The bandwidth can be considered a smoothing parameter, with larger bandwidths resulting in a smoother surface of estimates. The kernel refers to whether the bandwidth is constant across the study area (i.e., fixed kernel) or allowed to vary (i.e., variable kernel). In case of the former, a Gaussian function is used to weight all observations in a data set, but in the latter, a bi-square function is implemented in conjunction with the bandwidth that determines the number of observations to be used to estimate local parameters.

Unlike OLS regression that returns a single parameter estimate for each modeled relationship, GWR returns a set of estimates and standard errors that are associated with the location of each observation that can and should be mapped. The interpretation of the GWR estimates is identical to that used in OLS, except for the fact that there are far more estimates to be interpreted. Changes in the magnitude and direction of the estimates across a data set indicate that the relationships between variables in a model vary geographically (i.e., spatial heterogeneity). Measures of fit for GWR rely upon the Akaike Information Criterion (AIC) which provides a method to compare GWR models that use different mixes of independent variables to each other, as well as to the global OLS model. It should be noted that GWR presents several methodological challenges with regard to significance tests and analysis of variance and are the subject of ongoing research (Charlton and Fotheringham 2009).

Election data, and in particular, Italian election data, are well suited for analysis with GWR in light of the geographic dimensions to Italian politics. The next section provides a worked example of GWR, and examines recent support for the *Lega Nord* within one of its regional strongholds: Lombardy.

A Worked Example: The Resurgence of the *Lega Nord*

One of the unexpected results of the Italian general election of 2008 was the resurgence in popular support for the *Lega Nord*. Famous for its rhetoric in defense of local and regional interests in northern Italy, hostility to foreign immigrants, and governmental alignment with Silvio Berlusconi and his various parties (first *Forza Italia* and now PdL), support for the *Lega Nord* nearly doubled since the last election of 2006. GWR is used to examine in detail this recent change in electoral fortunes for the *Lega Nord* across the party's regional stronghold of Lombardy.

Lombardy is a good setting to examine the recent success of the *Lega Nord* with GWR because it is the recognized stronghold for the party, it is economically diverse, home to small, medium, and large scale enterprises, and serves as a beacon to new immigrants (Mainardi 1998). There are also a relatively large number of *comuni* (municipalities) across the region, which arguably contributes to administrative fragmentation. Figure 4.3 maps the 2008 electoral performance of the *Lega Nord* across Lombardy and highlights the geographic concentrations and variations in support for the party. At the national level, the *Lega Nord* obtained 8.3 percent of the vote in 2008, yet the party averaged over 25 percent within its regional stronghold. High levels of support for the *Lega* tend to be concentrated in the northern reaches of Lombardy, to the north of Bergamo and around Sondrio. Support for the *Lega Nord* decreases considerably across the south of the region, as well as in the urban centers of Lombardy (e.g., Milano, Pavia, Cremona, Brescia). The recent election also permits us to test some general hypotheses about issues important to supporters of the *Lega Nord*, and in particular, questions concerning immigration and immigrants.

Despite the adages that globalization and Europeanization are good and that many countries have been built upon the work of immigrants, in many parts of the world, recent immigrants tend to be viewed more as a threat than as an asset. Furthermore, there are several aspects to this perceived threat that immigrants pose that political parties and politicians integrate into their own platforms, campaigns and rhetoric. Perhaps the most common anti-immigrant argument is based upon economics. Following simple (but flawed) logic is the fear that immigrants take jobs away from local workers. Moreover, since recent immigrants are surely poor and willing to work for less, their employment undercuts wages. Paralleling this argument is concern and consternation over the provision and abuse of local social services to immigrants and especially undocumented immigrants.

Exacerbating such economic arguments against immigration and immigrants are those based on cultural fear and social insecurity. By way of example, criminal acts occur with similar relative frequency across all societies, but those committed or believed to be committed by immigrants tend to resonate more with local populations for several plausible reasons. For instance, local identities and traditions are often considered to be under threat or displaced by the arrival of immigrants and foreigners are seen as more likely than locals to violate accepted social norms as defined in criminal law.

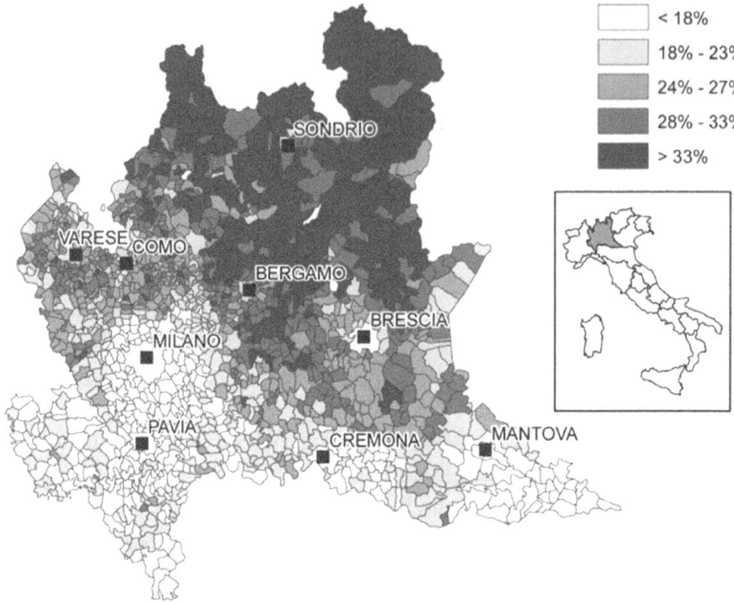

Figure 4.3 2008 Electoral Performance of the *Lega Nord* Across Lombardy
Source: Authors.

The *Lega Nord* is renowned for its continuing negative stance on immigration, as well as for its more positive but equally divisive views on political and economic issues such as devolution and federalism (Agnew 1995; Biorcio 1997; 1999; Diamanti 1993; 1996). In light of such political positions, and the discussion above about how immigration can be seen as a particularly important source of voter support for the *Lega*, three elementary but general hypotheses can be articulated to explore and possibly explain the resurgence in support for the *Lega Nord* using GWR:

- H_1: League support may be positively related to the presence of immigrants (immigrant threat)
- H_2: League support may be positively related to unemployment (immigrants take jobs)
- H_3: League support may be positively related to self-employed (defense of the local)

The above hypotheses draw from *Lega* rhetoric during recent election campaigns, and its historical anti-immigrant position. For instance, if indeed immigrants were taking jobs away, perceived as a threat to local traditions or involved in criminal activities, a positive relationship between *Lega* support and the relative size of the immigrant population is plausible. There is the possibility that such perceived threats occur in the local absence of migrants and are the result of second-hand

Table 4.1 Descriptive Statistics for Variables in Lombardy Data Set ($N = 1{,}546$)

Variable	Mean	Std. Dev.	Minimum	Maximum
Lega Nord 2008	25.8%	8.7%	5.3%	64.7%
Log (population)	3.71	0.54	1.81	6.41
Unemployed	4.4%	1.6%	0.0%	16.2%
Self-employed	17.0%	5.0%	4.6%	50.0%
Foreigners	6.7%	3.8%	0.0%	23.0%

Table 4.2 Baseline OLS Model Estimates for *Lega Nord* Support in Lombardy

Variable	Coefficient	Std. Error	t-value
Constant	0.480	0.022	
log (population)	-0.039	0.004	-8.65
Unemployed	-0.731	0.133	5.50
Self-employed	-0.144	0.047	-3.05
Foreigners	-0.272	0.058	-4.70

information (e.g., via media reports or anti-immigrant rhetoric and propaganda), a general fear of the unknown and not direct experience. Keeping this in mind, the hypotheses presented above are used primarily for illustrative purposes. Using aggregate data and GWR, the above arguments as they relate to the recent success of the *Lega Nord* are tested.

The following analyses are based on data obtained from the Cattaneo Institute in Bologna and the Italian national statistics bureau ISTAT for the 1,546 *comuni* in Lombardy. Descriptive statistics for the relevant variables are provided in Table 4.1.

Across Lombardy, the mean level of unemployment is relatively low, and average levels of self-employed and foreigners are moderately high, especially when compared to the rest of Italy. Each of the variables however exhibits a considerable amount of variation across the region. Using the above variables, a baseline, global OLS regression is estimated. Results from the OLS are reported in Table 4.2. The goodness of fit measures indicate that the model explains a relatively small amount of the total variance (i.e., $R^2 = 0.09$). Despite this relatively small value, parameter estimates from the model are statistically significant. However, the direction of the relationships between support for the *Lega Nord* and unemployment, self-employment and the number of foreigners are opposite of what was hypothesized. This may be the result of model (or theoretical) misspecification and merits further investigation.

Spatial diagnostic tests (unreported) indicate that a spatial lag model is preferred over a spatial error model. Estimates for the spatial lag model are reported in Table 4.3. As a measure of fit in the spatial lag model, a pseudo-R^2 index is calculated that compares the predicted variance to the observed variance in the model (see Anselin 1992). This value is 0.77, but perhaps more interesting is

Table 4.3 Spatial Lag Model Estimates for *Lega Nord* Support in Lombardy

Variable	Coefficient	Std. Error	t-value
Constant	0.092	0.013	
spatial lag (*Wy*)	0.858	0.014	62.17
log (population)	-0.010	0.002	-4.32
Unemployed	-0.334	0.070	-4.95
Self-employed	0.026	0.024	1.09
Foreigners	-0.103	0.029	-3.48

a comparison of the parameter estimates between the spatial lag and baseline OLS models. Specifically, the estimates for population, unemployed and foreigners remain significant but decrease in magnitude in the spatial lag model, and the self-employed variable becomes positive, though it is no longer statistically significant. As with the OLS model, the direction of the predicted relationships are opposite of what was expected. The spatial lag of *y* is remarkably significant and indicates that the spatial structure within the data cannot be ignored.

Recall that the spatial lag model assumes that parameter estimates are homogeneous or stable across the data set. To test this assumption and to explore further the spatial structure of the Lombardy data set, geographically weighted regression (GWR) is used. This analysis was carried out with the ArcGIS 9.3. An adaptive kernel was used, and bandwidth was selected on the basis of minimizing the AIC value (see Fotheringham et al. 2002 for details). The procedure for determining bandwidth is automated and involves assessing difference between observed and fitted values, as well as the complexity of the model. As noted earlier, estimates for each and every observation are returned by GWR, and are best visualized through maps.

The first map from the GWR is for the standardized residuals (Figure 4.4). Darker shades indicate areas where the model under-predicted *Lega* support and lighter shades indicate areas where the model over-estimated *Lega* support. This map shows clearly that support for the *Lega Nord* is underestimated consistently in the urban centers of Lombardy (e.g., Milano, Brescia) Moreover, in some areas, such as around Milano and Brescia, and to the southeast of Cremona, there are pockets or clusters of residuals of similar magnitude. Such patterns reinforce the need to consider carefully the geography of elections, and even call into question the utility of a single parameter estimate for a region such as Lombardy.

Maps of the local coefficients for each independent variable (except population) clearly show the spatial structure that underlies the data, as well as the spatial heterogeneity in the nature of relationships for a given variable, as well as between variables. It is useful to keep in mind the actual estimates obtained from both the OLS (Table 4.2) and spatial lag (Table 4.3) models for comparison and as a starting point for interpretation. Recall that the estimates for unemployment returned by the OLS and spatial lag models were, -0.73, and, -0.33, respectively. The GWR map (i.e., surface) of local unemployment coefficients shows where in Lombardy such

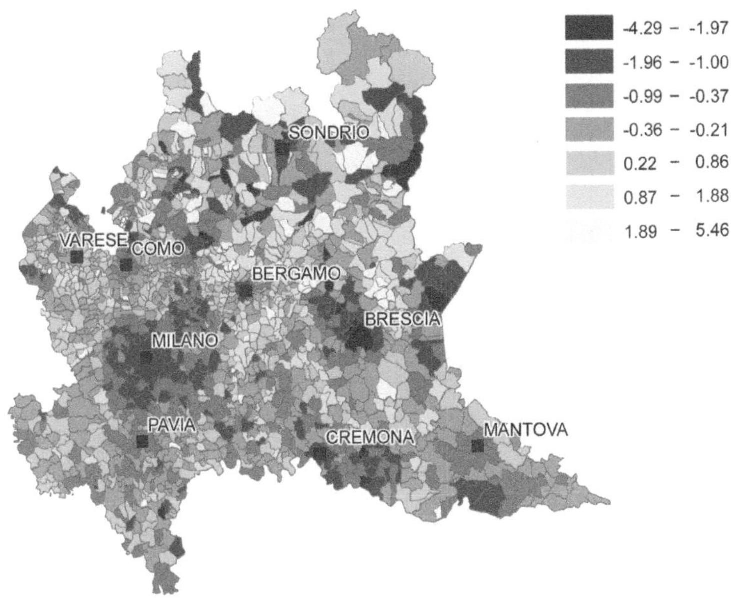

Figure 4.4 Geographically Weighted Regression Residuals Across Lombardy
Source: Authors.

estimates, or in other words, the relationship between support for the *Lega Nord* and unemployment, are both positive and negative (Figure 4.5). For instance, the relationship is positive in areas to the north of Brescia, the west of Sondrio and the east of Mantova. Conversely, there is a strong inverse relationship between unemployment and *Lega Nord* support to the east of Milano. To put into context the geographic variation of this relationship, the local GWR estimates vary from a minimum of -3.17 around Milano to a maximum of +0.76 in a few of the *comuni* in the northern and eastern extremities of Lombardy. Hence, in some areas the results from the GWR support our initial hypotheses regarding the linkage between unemployment and support for the *Lega*, yet in other areas our hypothesis is rejected.

The map of GWR estimates for the association between foreigners and *Lega* support also reveals some areas where the relationship is negative and one area in particular where it is positive (Figure 4.6). With regard to the latter, the cluster to the east of Milano and south of Bergamo contains GWR estimates between +0.37 and +0.58. For comparison, OLS returned an estimate of -0.27 and the spatial lag model returned an estimate of -0.10. Though this result (i.e., cluster of GWR estimates) supports our hypothesis that the presence of foreigners is linked to higher levels of support for the *Lega Nord*, this is only the case in a certain area of Lombardy. The map also identifies areas of Lombardy, and in particular to the east and west of Sondrio, where the association between number of foreigners and *Lega* support is negative and more in line with the estimates returned from

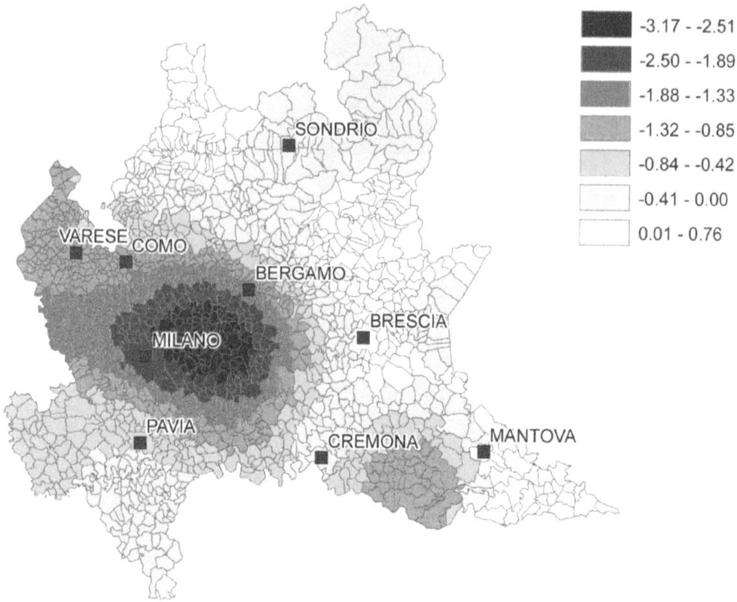

Figure 4.5 Geographically Weighted Regression Map of Local Unemployment Coefficients Across Lombardy
Source: Authors.

the OLS and spatial lag models. Such discrepancies in both parameter estimates across models, as well as returned by GWR, may be a result of theoretical or model misspecification. Regardless the cause of such differences, such discrepancies and geographic variations merit careful interpretation and consideration.

The map of local coefficients for the relationship between the self-employed and *Lega* support again shows how the association is positive in some areas of Lombardy, but negative in others (Figure 4.7). The most conspicuous cluster of positive estimates is situated between Milano and Bergamo, in the heart of Lombardy. The magnitude of the association between self-employment and *Lega Nord* support tends to decrease as distance increases from this central cluster, with the exception of a local concentration of positive local estimates around Mantova. The pockets of negative estimates appear on the western, northern and eastern borders of Lombardy. Like the previous maps, the results from the GWR highlight the geographic instability in the relationship, in this case, between the self-employed and *Lega* support. Again, depending upon where GWR coefficients are obtained, our hypothesis could either be accepted or rejected at the local level.

The above worked example illustrates the potential that GWR has within the context of electoral studies and highlights how it differs from both OLS and spatial econometric techniques. The exact choice of regression technique to use when examining a given election is perhaps less clear, and is likely to reflect exposure and

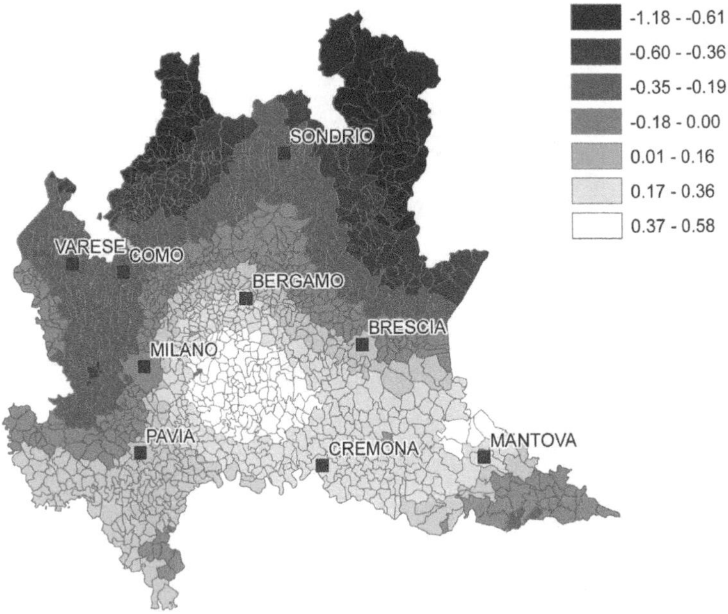

Figure 4.6 Geographically Weighted Regression for the Association Between Foreigners and *Lega Nord* Support

Source: Authors.

familiarity with geographic concepts, methods and issues (e.g., scale, spatial analysis, spatial autocorrelation) in general. For instance, OLS is used and understood widely, which is arguably its greatest strength. Despite recent innovations and diffusion of geo-statistical and spatial analytic techniques, the simplicity and accessibility of OLS is its greatest advantage. As noted earlier, however, is the fact that many data sets are spatially biased, which in turn can bias or influence the results from OLS.

Spatial econometric methods, such as the spatial lag model, introduce a level of complexity to regression that in some cases may be necessary in order to obtain unbiased and efficient parameter estimates. The trade-off when using geo-statistical techniques versus OLS is that much more additional information about a particular data set must be obtained or generated. For instance, the geographic locations of each observation, and the geographic relations between observations, need to conceptualized and recorded. Though this is a trivial exercise within the context of geographic information systems (GIS), it is a non-trivial issue for the uninitiated.

What is more is that the interpretation of spatial econometric models is not necessarily straightforward, but depends upon whether geography (i.e., the spatial bias within a data set) is conceived and understood as merely a nuisance that needs to be controlled or a substantive effect that needs to be explained

Furthermore, conceptualizing geography as simply an additional parameter within a regression model is in many ways problematic and unsatisfying because geography

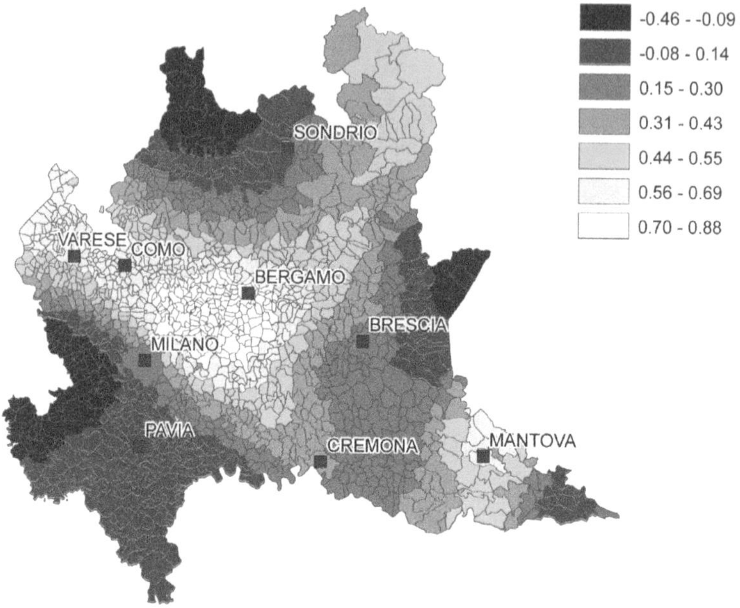

Figure 4.7 Map of Local Coefficients for the Relationship Between the Self-Employed and *Lega Nord* Support

Source: Authors.

is not necessarily an additive effect; rather, geography contextualizes the variables and relations of interest differently in different places. In other words, unemployment in and around Milano may have different connotations and impacts upon the local population when compared to unemployment in a more rural setting such as Sondrio.

As an exploratory technique, GWR perhaps provides the most (local) insights into the geographic nature of relationships in aggregate data sets. In addition to the methodological complexity and technical requirements of GWR, one of the greatest trade-offs is the simplicity in the interpretation of results. In the case of evaluating and assessing relationships with OLS and spatial econometric techniques there is only a single coefficient to interpret. GWR returns a coefficient for each and every observation. Mapping such estimates certainly facilitates interpretation, but there is also the presumption that such maps as well as the GWR results themselves can be interpreted and understood. Furthermore, though GWR can provide interesting if not valuable insights into the geographic instability of regression coefficients, assessing statistical significance remains problematic, and perhaps more importantly, GWR has the potential to introduce a prejudice for the local. Just as OLS and spatial econometric methods can be criticized for returning single or global parameter estimates for a data set, are we missing the forest for the trees with GWR? Such questions and trade-offs surrounding the use and application of OLS, spatial econometrics and GWR are more about how and why geography matters to electoral studies, and are less about methodology per se.

Conclusion

The spatial regression techniques discussed and implemented above hold considerable potential for electoral studies and analysis. In the absence of survey data, or used in conjunction with such individual-level data, spatial regression can complement and extend our understanding of voter attitudes and behavior in several ways. For instance, such methods can be used to refine and articulate research questions, and in particular, questions concerned with the influence of geographic context upon political behavior and electoral change. Analyses such as that presented above can also be used to guide subsequent studies, direct fieldwork or even inform sampling strategies for voter surveys.

What is clear is that the spatial regression techniques provide a useful and accessible means to evaluate and assess how spatial autocorrelation and spatial heterogeneity are possibly implicated in models of voter behavior, and more generally, the voting choice itself. In the worked example that modeled 2008 support for the *Lega Nord*, it is clear that when the spatial structure underlying the data are disregarded and ignored, the parameter estimates obtained via OLS are biased. The introduction of a spatial lag term to control for spatial autocorrelation (i.e., Wy) in the spatial lag model provides insights into the nature and degree of parameter estimate bias, as well as a potential control for the underlying spatial structure of the data.

Geographically weighted regression (GWR) was then introduced to extend further the geostatistical modeling of recent support for the *Lega Nord*. The added value of the GWR approach, over both the standard OLS and spatial lag models estimated, is its ability to evaluate and show how parameter estimates vary geographically and locally. In other words, GWR recognizes that the relationships between variables may not be the same, or homogeneous, across a given data set. By obtaining local parameter estimates, such geographic variation provides more detailed and nuanced insights into the relationships between, for example, *Lega Nord* support, socio-economic indicators and geography. At the very least, the spatial regression methods outlined highlight the need to consider geography when estimating models with georeferenced data, and may in fact promote a more explicit treatment of geography with regard to theories of political behavior.

Finally, it should be noted that the choice of how such geographic relationships between observations are captured by GWR or by the spatial weights matrix, **W**, in the case of the spatial econometrics, is non-trivial and can have implications for the interpretation of geographic effects. In this regard, it is important to determine how and why geography matters. This last point is a theoretical issue, and not a methodological or technical one. For example, what processes and mechanisms account for *Lega Nord* support in one *comune* to be related to those in neighboring *comuni*? How consistent is this relationship within a given province or across Lombardy? What accounts for the variation in the nature of the relationships between support for the *Lega Nord* and unemployment, self-employment and immigrant presence across Lombardy? As this example illustrates, spatial analysis, spatial econometrics and GWR hold a considerable amount of promise with respect to answering and extending research questions in electoral studies, as well as articulating and informing future directions for research in political science and allied disciplines.

References

Agnew, J. 1995. The rhetoric of regionalism: The Northern League in Italian politics. *Transactions of the Institute of British Geographers* 20: 156–72.

Anselin, L. 1995a. Local indicators of spatial association – LISA. *Geographical Analysis* 27: 93–115.

Anselin, L. 1995b. *SpaceStat, version 1.80 user's guide*. Morgantown, WV: Regional Research Institute.

Anselin, L. 1988. *Spatial Econometrics: Methods and Models*. Dordrecht: Kluwer Academic.

Biorcio, R. 1999. La Lega nord e la transizione italiana. *Rivista italiana di scienza politica* 22: 55–88.

Biorcio, 1997. La *Padania promessa. La storia, le idee e la logica d'azione della Lega Nord.* Milano: Il Saggiatore.

Brunsdon, C., A. Fotheringham and M. Charlton. 1996. Geographically weighted regression: A method for exploring spatial non-stationarity. *Geographical Analysis* 28: 281–98.

Cartocci, R. 1990. *Elettori in Italia.* Bologna: Il Mulino.

Charlton, M. and A. Fotheringham. 2009. *Geographically Weighted Regression.* National Centre for Geo-computation White Paper. Maynooth, Ireland: National University of Ireland.

Cliff, A. and J. Ord. 1981. *Spatial Processes: Models and Applications.* London: Pion.

Diamanti, I. 1996. *Il male del Nord.* Roma: Donzelli.

Diamanti, I. 1993. *La Lega.* Roma: Donzelli.

Fotheringham, A., C. Brunsdon and M. Charlton. 2002. *Geographically Weighted Regression: The Analysis of Spatially Varying Relationships.* Chichester: John Wiley & Sons.

Galli, G. and A. Prandi. 1970. *Patterns of Political Participation in Italy.* New Haven, CT: Yale University Press.

Mainardi, R. 1998. *Il Nord e la Padania. L'Italia delle regioni*. Milano: Bruno Mondadori.

O'Sullivan, D. and D. Unwin. 2003. *Geographic Information Analysis*. Hoboken, NJ: John Wiley & Sons.

Shin, M. and J. Agnew. 2008. *Berlusconi's Italy: Mapping Contemporary Italian Politics.* Philadelphia, PA: Temple University Press.

Chapter 5

Elections and Cultural Political Economy: The Political Geography of the *Bloque Nacionalista Galego* in the Galicia Autonomous Community

Erinn P. Nicley

In the autumn of 1993, Spanish media reporting cast a spotlight on the rise of the Galician Nationalist Bloc (BNG) in upcoming regional parliamentary elections in the northwest province of Galicia. The BNG coalition and its charismatic presidential candidate, Xosé Manuel Beiras, appeared to be gaining steadily against other left-wing political forces in the traditionally conservative region. After an intense electoral campaign, the ruling conservative People's Party (PP) regained an absolute majority control over parliament and the presidency. The Socialist Party of Galicia (PSdeG) – the regional affiliate of the Spanish Workers' Socialist Party (PSOE) – retained its second-place opposition status with nearly 24 percent of the vote and 19 parliamentary seats. And the left-wing, regionalist BNG posted a third-place showing with 18 percent and 13 seats in the 75-member regional parliament (Parlamento de Galicia 2010).

The BNG finish as a third-place, regionalist political force in 1993, at first glance, is an unusual foundation for a book chapter on new perspectives on electoral geography. The 1993 Galician elections had not fundamentally changed the electoral landscape. True, the BNG had more than doubled both voter support and parliamentary seats won during the 1989 parliamentary election and had greatly exceeded its 1982 electoral maiden voyage, in which the party received only 4 percent and one parliamentary seat. However, the overwhelmingly conservative Galician political climate continued to prevail despite the internecine political turmoil transforming progressive political forces in the race for second place.

The importance of the 1993 Galician regional elections is found not in terms of control over the Galician political administration, but rather for what the patterns of BNG support can tell about the mutual relationship between the political economy of European structural integration, regionalism, and the place-specific struggles that constitute the broader political and economic milieu. The rise and decline of the BNG during a series of regional parliamentary elections from 1985–2009 reflects a regionalist party and a broader nationalist movement caught in the maelstrom of political and economic forces struggling to redefine and reorganize the geography of political power and cultural identity across multiple scales.

This chapter approaches the political geography of elections through a cultural political economy perspective (CPE). This CPE approach combines political economy, cultural studies, and semiotics to reposition electoral studies as a window onto the broader connection between macro-scale forces and the contingent, culturally specific organization of political and economic relations. A CPE perspective allows us to approach elections and electoral processes as a political moment in which the grounded electoral struggles that occur within particular places simultaneously are influenced by and play a part in reconstructing from within the socio-spatial structure that conditions political, economic, and social life.

In this vein, I analyze the political geography of the rise and decline of the BNG coalition during the 1985–2005 Galician regional parliamentary elections. For brevity, I focus on BNG electoral success during the 1993 and 1997 elections to provide a unique, if imperfect, window into the place-specific political struggles that have surrounded Galician and Spanish integration into the European political-economic structure. The geographic and historical patterns of BNG electoral support spotlight the political-economic conditions that created the political space for the BNG political platform to attract renewed support among Galician electorate. The political geography of BNG support demonstrates that BNG electoral policy has been very responsive to the different political-economic position of particular Galician localities. At the same time, the BNG has been able to discursively incorporate existing social and cultural conditions within distinct localities to construct regionalist narratives that attempt to reposition the BNG and the broader Galician territory as a political force within the broader European and Spanish political economy.

A Political Geography of Elections

This chapter begins with a question: What would or *could* a political geography of elections look like? I purposefully adopt the language of 'political geography' to emphasize the embeddedness of elections within a more expansive view of the political. A political geography of elections begins with the simple recognition that the object of study extends both before and after actual elections. As the chapters in this volume attest, such a political geography of elections affords an opportunity to apply critical and post-structural perspectives towards many existing electoral geography themes. For example, geographers are well-placed to discuss the theoretical foundations and development of formal and informal electoral institutions, political party organization and operations, campaign finance, electoral participation (and non-participation), media election coverage, the gendered, racialized, and sexualized geographies of elections, and post-structural perspectives that see elections as embodied political performances that produce particular societal discourses and effects. The shifted focus onto critical and post-structural election studies likewise suggests that greater use of qualitative methods may help uncover new understandings of electoral processes as grounded, contingent, and embodied processes deeply intertwined with the broader societal milieu.

In this chapter, I apply a cultural political economy perspective towards the political geography of elections. Cultural political economy has emerged in recent years as a promising theoretical approach that can bridge the divide between critical political economy and the post-structural concern with identity politics, discourse, and political performance (Jessop and Oosterlynck 2008; Jessop and Sum 2001). To begin, I want to delimit my use of the term 'political economy'. I specifically refer to the processes of capitalist regulation that develop and reproduce a particular political and economic regime. Regulationist thought in recent years has developed a concept of *societal regulation* to capture the diverse array of economic and extra-economic institutions and relations through which capitalist political economy is reproduced (Jessop and Sum 2007). This has opened new avenues to seeing capitalist regulation of the political economy in unexpected places seemingly far removed from the macro-economic worlds of financial regulation, market institutions, and economic class relations. For this study, we can define capitalist regulation as a disciplinary regime that blends together both economic and extra-economic social institutions and relations to normalize the capitalist political and economic structure and thus secure continued capital accumulation (Jessop and Sum 2007).

A cultural political economy perspective examines the culturally mediated institutions and relations through which the political economy of capitalist regulation is produced on the ground (Jessop and Sum 2007; Sayer 2001). Cultural political economy offers a compromise between the monolithic class-centered vision of structural political economy and the post-structural vision of multiple, situated experiences of capitalism fractured by race, gender, and other social identities (Gibson-Graham 1996). That is, cultural political economy stands apart by seeing capitalist regulation as taking shape through the contingent intersections of economic and identity politics that constitute everyday experience in particular places. The political and economic structure in which localities are embedded is itself constructed through the culturally distinct political relations that mediate broader social forces within particular places. The contingent forms of capitalist regulation that emerge through such grounded political relations establish distinct geographic patterns of political and economic power across different scales and among particular places in relatively stable but impermanent configurations (Jessop and Sum 2007).

From this perspective, the political-economic structure is socially produced through the culturally distinct political institutions and relationships that build from within hegemonic patterns of capitalist regulation in particular places. However, cultural political economy places limits on the range of culturally mediated political and economic discourses that may form within the broader capitalist context. The cultural political economy takes root through an evolutionary process that privileges certain culturally distinct political and economic possibilities – what Jessop and Oosterlynck (2008) call economic imaginaries – that also tend to best support capitalist accumulation. Thus, cultural political economy involves a reflexive process of variation, selection, and retention in which competing economic discourses circulate within capitalist society (Jessop and Sum 2007). The selection of certain economic imaginaries privileges a culturally distinct set of narratives, semiotic meanings, and material practices that gain traction in particular places to contingently reproduce the broader conditions of capitalist reproduction. The socio-spatial structure of the political economy arises through repeated instances of definition, negotiation, reproduction, and transformation as it both responds to but also reproduces the culturally mediated conditions of capitalist regulation and political economy (Jessop and Oosterlynck 2008).

This cursory overview of cultural political economy reflects the abstract state of current theory on the actual operation of cultural political economy at the intersection of semiosis, discourse, and critical political economy. Geographer Martin Jones has called recently for greater attention to the embodiment of cultural political economy as a political performance (Jones 2008). The cultural political economy is reproduced on a daily basis through particular institutions, processes, and political actors. In this sense, we can grasp the continued reproduction of cultural political-economic conditions best through research focused on specific political institutions, events, narratives, and relations that play a role in maintaining (or transforming) the dominant conditions of capitalism in particular places. As Jones (2008, 396) notes, the 'centrality of the meaning and spatialities of politics itself [becomes] the primary focus of conflict and debate'. In effect, the focus on real-world political moments – like democratic elections – becomes a window into the economic and extra-economic reproduction of the cultural political economy of capitalist regulation in everyday life.

Elections are a distinct political moment within a society. The electoral process reflects the embodied performances and struggles among political and economic forces to define and put into practice a political and economic program. From this perspective, elections extend beyond the limited experience of political parties on election-day to include multiple social forces that take part in and influence the electoral process, e.g., political parties, social movements, constituent communities, media producers, financial interests, legal institutions. This extended view suggests that elections in many ways reflect the prevailing political-economic conditions within a society. Of course, political movements frequently reflect the political economy through their internal composition, their programs, and their constituent bases of support.

Political geographers have highlighted from different perspectives the geographic connections between political economy, place, and electoral politics (Agnew 1997, 2000; Clayton 2002; Flint 1996, 2001; Keating 2001; Marden 1997; Shin 2001). For such writers, a regional and place-based political economy perspective has helped explain the geographic variation among electoral processes and electoral outcomes across a range of political, regionalist, and ethnic nationalist movements worldwide. The role of political parties and institutions in the social construction of political spaces has been central to many of these arguments. For example, Flint (2001) draws upon regional and place-based political-economic conditions as the context upon which the Nazi party strategically tailored electoral narratives and discourses to generate regional appeal in pre-war Germany. For such studies, the political economy conditions the role of political discourse in creating the distinct political geographies of elections. At the same time, there is a recognition that political economy is not only a shaping influence, but also is actively being produced through place-specific political struggles to define, put into practice, and institutionalize particular institutions and relations (i.e., political economy). Elections afford political parties an important, albeit de-centered, opportunity to reconstruct the geographic 'horizons' of political and economic discourse within the bounded concept of particular regions, places, scales, and networks (Agnew 1997).

A cultural political economy perspective advances the political geography focus on the construction of electoral political space with exploration of how elections can serve to reproduce culturally mediated modes of capitalist regulation in particular places. The electoral process in part is a political consequence of structural political and economic forces that shape the field of political issues, political movements, and voter mobilizations in a particular time and place. However, we can use a cultural political economy perspective to unearth the construction of political-economic conditions through the distinct social institutions, relations, and narratives that constitute political and economic life on the ground (MacLeod 2001). It is these everyday geographies that both mediate and actively *produce* the political and economic conditions that both shape and emerge through electoral outcomes.

Elections can be seen as an opening for renegotiation of the dominant cultural political economy. Elections are a moment in which the dominant political-economic 'imaginaries' are actively open to debate among political forces with alternative visions for the organization of political-economic space. In other words, elections are a snapshot of the culturally mediated political-economic fault lines within a society. Yet, elections also permit the reconstruction of a new cultural political economy through the selection of new political and economic narratives and practices that ground capitalism within particular places and regions. Political parties and other electoral forces selectively draw upon existing identities and cultural themes as the foundation for rebuilding the political-economic position of the society within the broader capitalist structure. In this way, elections are a *performance* of the cultural political economy in which competing forces struggle to arrange, reproduce, and transform the social and spatial patterns of political and economic life through the culturally distinct field of relations that comprises everyday life.

These tentative thoughts on a cultural political economy perspective on the political geography of elections return us to the initial question: What can a political geography of elections look like? The possibilities that I am beginning to articulate in this chapter highlight that elections are at once micro-scale political events and bound recursively into broader political and economic forces. The call to unite the macro- and micro-worlds of electoral geography is well-worn territory (e.g., Flint 1996). The cultural political economy perspective takes us further with a focus on the place-specific identity politics that mediate, give meaning and material form to capitalist political economy in time and space. Elections are a key moment in the contingent reproduction of these culturally mediated political-economic visions that define the position of particular places within the broader capitalist structure.

We can consider a cultural political economy perspective to develop a roadmap for studying the connections between capitalist regulation and the political geography of elections. This roadmap consists of three parts:

1. examination of the political-economic context shaping the geography of electoral outcomes;
2. examination of political campaign narratives and how political forces select and discursively draw upon particular cultural identities and social relations to promote particular political-economic imaginaries through the electoral process; and
3. examination of how elections and electoral outcomes renegotiate the cultural political economy of capitalist regulation through place and scale-specific political processes that feed into the reconstruction of the capitalist structure 'from within'.

In the remaining sections of the chapter, I explore the connections between the cultural political economy of capitalist regulation and the political geography of regional elections in the Galician autonomous community in northwest Spain. The roadmap described above structures the conversation to analyze the geography of BNG political support in terms of political-economic changes that came with Spanish adhesion to the European Union, the political response that the BNG offered to the social dislocations experienced in particular Galician localities, and the transformation of the Galician regional political-economic position within the European regional structure through the political presence of the BNG and its regionalist political discourse.

The Political Geography of Regional Elections in Galicia

The rise of the Galician Nationalist Bloc (BNG) as a central political force in Galician political life in the past 30 years emphasizes the powerful influence of place-based political economy upon the political geography of elections. The historical regional context of Galician political, economic and cultural marginalization has influenced Galician politics for over a century. The varied

Galician regionalist and nationalist movements from the 1840s to the present have conflated Galician cultural identity with traditional *genres de vie* rooted in the maritime fishing, animal husbandry, and agricultural sectors (Sharif 2006). In this sense, the BNG story is also the story of Spanish adhesion to the European Union and its predecessor organizations during the 1980s and 1990s. The renegotiation of the Galician political economy through the turbulent social processes of European–Spanish integration became a focal point for Galician regionalist politics discursively committed to defending Galician cultural and political-economic heritage. From EU–Spanish adhesion in 1986 to the political climax and *dénouement* in the early 1990s, the political geography of successive regional parliamentary elections reflected the blended economic conditions and identity politics that occupied the hearts and minds of the Galician population.

The *Bloque Nacionalista Galego* was a marginal regionalist political force in the early years of post-Franco Spain (1978–present). Galician regionalist political forces proliferated among both the political right and a deeply fragmented political left eager to reclaim political space lost during the Franco dictatorship. The BNG was no exception when it formed and first participated in the 1982 regional election. The diverse positions on the question of independence vs. varied levels of self-determination and autonomy as provided in the 1978 Spanish constitution limited the success of left-leaning regionalist political forces at the polls (Botella 1989). Indeed, Galician regionalist electoral politics in the early 1980s suggested that the center-right Galician Coalition (*Coalición Galega, CG*) was going to set the Galician regionalist agenda from its electoral stronghold in the rural, interior Ourense and Lugo provinces of eastern Galicia. The meritable 13 percent support for the Galician Coalition in the 1985 regional parliamentary elections stands in contrast to the BNG, who gained only 4 percent support from radicalized centers on the Galician coast (Vilas Nogueira 1992).

A brief four years later, the BNG rose in the 1989 regional elections on a political trajectory that ultimately positioned the BNG as the hegemonic regionalist force on the Galician political landscape for the next 20 years. In subsequent regional elections, the BNG continued to demonstrate its capability to attract votes as a major opposition party: 8 percent (5 seats) in 1989, 18.4 percent (13 seats) in 1993, and 24.8 percent (18 seats) in 1997. By the 2001 regional election, the BNG surpassed the socialist PSdeG-PSOE as the main opposition party with 22.6 percent (17 seats) voter support. Mixed BNG success in 2005 greatly reduced voter support to 18.7 percent (13 seats), but gave the BNG the Galician vice presidency in a coalition government with the socialists. The BNG suffered a stark decline in 2009 regional elections, receiving only 16 percent support (12 seats), compared to 25 seats won a resurgent socialist PSdeG-PSOE (all data from Parlamento de Galicia 2010).

The BNG success through the 1990s and early 2000s and its uncertain future today spotlight the question of why and how a far-left regionalist political force was able to gain so much political ground in a territory renowned for its traditionalist, conservative politics (Vilas Nogueira 1992). Scholars frequently point to the proximate causes of BNG growth during these years. The incorporation of moderate regionalist political parties into the BNG coalition, the decline of the state-range PSdeG-PSOE, and the mobilization of new and idle voters have all played their part in BNG success (Barreiro Rivas 2003; Márquez Cruz 2004; Portas 2010; Vilas Nogueira 1992). However, closer examination of the political geography of BNG electoral support during the critical 1993 and 1997 regional parliamentary elections suggests that the Galician electorate increasingly turned to the BNG as a place-specific political response to the European political-economic integration and the appeal of the BNG as self-proclaimed protector of Galician culture and position with the emerging European system (Casal and Ventoso 1994).

In this section, I analyze the political geography of BNG electoral support in a series of regional parliamentary elections from 1985–2005. Exploratory data analysis of BNG percent vote received in each election using ArcMap 9.2 Gi* hotspot analysis highlights a complex political geography punctuated by multiple spatial clusters of significantly high and low BNG support. The BNG electoral map series reveals spatial and longitudinal patterns of changing BNG support among 315 Galician municipal townships that constitute the principal electoral candidate districts throughout the region. Following an initial review of the entire 1985–2005 BNG electoral series, I focus on the 1993 and 1997 regional elections to examine the relationship between BNG electoral clusters and the reconstruction of the Galician cultural political economy through the electoral process.

The generalized BNG growth from the 1985–2005 regional elections demonstrates quite graphically the consolidation of regionalist political forces and the ideological shift towards a more moderate, social-democratic discourse from 1987 onwards (Barreiro Rivas 2003). From 1987, the BNG formalized a new strategy of moderation of its radical discourse, a process that began with the acceptance of the Spanish constitution in 1985. In place of the historic rejection of the Spanish state, a new BNG discourse emphasized maximization of Galician autonomy with social-progressive themes and instrumental concerns involving Galician political, economic, and cultural affairs (BNG 2010). By 1989, the BNG reoriented its program towards protection of Galician political, economic, and cultural interests within the complex political economy of the Spanish state and European community. Throughout the early 1990s, the BNG increasingly attempted to attract moderate voters in the traditional socialist-leaning interior townships of A Coruna province, Pontevedra province and, with notable success, the former Galician Coalition strongholds in the Ourense province of southeastern Galicia. The early 1990s Galician political focus on economic recession, debate regarding approval of the Maastricht Treaty on European Union, and the growing debilitation of the Spanish socialist (PSOE) government both nationally and within Galicia were central to BNG success (Casal and Ventoso 1994).

The 1993 and 1997 regional elections maps highlight three visible and statistical trends in the data (Figure 5.1 and 5.2). First, the BNG from 1985–2005 has demonstrated a general expansion of political appeal throughout Galician

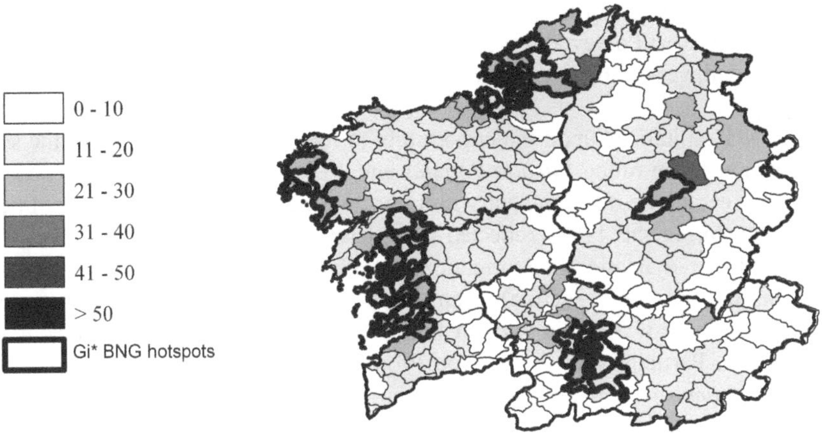

Figure 5.1 BNG Election Support Map, 1993 (%)

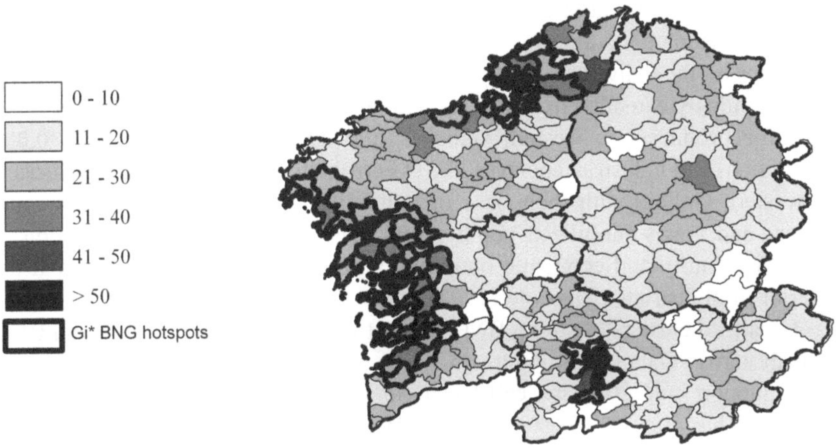

Figure 5.2 BNG Election Support Map, 1997 (%)

territory. The growth of BNG support clearly remains strongest among the western A Coruna and Pontevedra provinces, but BNG expansion into the rural, conservative Lugo and Ourense provinces to the east is notable. Second, the BNG simultaneously developed deeper political support among its core constituent townships and cities along the heavily populated Galician west coast. Third, the 1993 and 1997 regional elections most distinctly highlight the consolidation of three distinct BNG support clusters centered on:

a. northwest A Coruna province;
b. southwest Pontevedra province; and
c. southwest Ourense province.

The BNG capitalized upon the opportunity to reposition its discourse as the sole opposition to the ruling *Partido Popular* and positioning itself as the defender of Galician economic development within the European Union. The BNG political program during the 1990s reflected a moderate, inclusive political project that reflected the dynamic transformations of Galician political economy during the period (see next section). Campaign slogans motivated Galician voters with narratives of 'Galicia without limits' (*Galiza sen limites*), 'Open Yourself to Galicia. Participate in a Common Project' (*Abre-te a Galiza. Participa nun proxecto común*) and 'The Galicia that you want. Together we can' (*A Galiza que queres. Entre todos podemos*) (BNG 2010).

The BNG political program yielded results as BNG support advanced from its traditional western enclaves into central and eastern Galicia in the 1993 and 1997 regional elections. This growth can be seen as a response to the concurrent development and intensification of European-led political and economic restructuring and the failure of Spanish-range political forces (and their regional allies) to adequately address growing Galician concerns over the effect of these processes on the daily lives of key constituents in the Galician community. BNG support quickly extended from its base among industrial workers, agricultural and maritime producers in western Galicia's mid-sized townships (10,000–30,000 population) and industrial enclaves like O Ferrol towards the larger urban districts of western Galicia with their base of culturally conscious liberal professional classes and newly mobilized youth voters concerned with the deterioration of Galician identity and economic opportunity both along the urban coast and the interior rural districts (Aizpeolea 1997; Bauer 1992; Rivas 1997).

Localized clusters of strong BNG support further punctuate the regional BNG growth across Galicia. As Figures 5.1 and 5.2 indicate, geographic patterns of BNG support among what I call the *A Coruna cluster*, the *Pontevedra cluster*, and the *Ourense cluster* suggest that BNG political strength in the 1990s reflects particular place-specific contextual factors beyond mere demographic variability among the Galician electorate. In a previous, unpublished study, I have demonstrated that BNG support among these three geographic clusters are in part a response to place-specific economic structures and underlying political-economic transformations within each cluster (Nicley 2002). This structural-ecological perspective suggested that BNG support had concentrated among townships where the European integration process had introduced a series of structural and sociological changes effecting traditional Galician economic sectors and production complexes. Structural changes to primary sector maritime production centered in the *Pontevedra cluster*, traditional shipbuilding and steel production in the *A Coruna cluster*, and a mix of dairy production, timber and lumber production, and environmental pollution concerns in the *Ourense cluster* correlated statistically with the place-specific geographies of higher-than-normal BNG support in each region (Nicley 2002).

These BNG spatial patterns corresponding to the BNG peak success in the 1993/1997 regional elections compel a closer examination of European integration and Spanish state policy as factors driving BNG growth and stabilization during the past 20 years. In the following two sections, I explore the Galician economic structure and the grounded experiences of European political-economic reform that accompanied the Spanish and Galician adhesion to European Community/European Union policies, regulations, and institutions. The chapter concludes with consideration of how the transformation of the Galician economic structure during the 1990s established a political opportunity for the BNG to reconstruct the cultural political economy of EU–Spanish–Galician relations through the elections process.

European Integration of the Galician Political Economy (1985–2005)

To make sense of the Galician Nationalist Bloc growth during the 1990s, we first must understand the transformations to the Galician economic structure experienced during the late 1980s–1990s. Territorially, Galicia is a minor part of the Spanish state – slightly smaller in area than the US state of Maryland and with a stable, modest population of 2.8 million inhabitants. The Galician position within the Spanish state since the mid-nineteenth century reflects the longstanding tension between structural underdevelopment and limited Spanish state programs to promote economic diversification (Beiras 1994). Despite considerable development of both secondary industry and tertiary services in the past three decades, traditional primary sector economic activities (i.e., fishing, dairy, agriculture, forestry) remain stalwarts of the Galician economy and social structure (Figure 5.3). Galician political economy characteristically involves a simultaneous mixture of subsistence and market-based production. Occupational identities are blurred by the carrying-out of multiple primary, secondary, and tertiary sector economic activities as part of daily life among Galician households (Baylina and García-Ramón 1998; Roseman 1999). Traditional male economic activities, such as industrial labor, construction, forestry and fishing, are frequently coupled with traditional female economic activities, including subsistence agriculture and dairy production, but also with clam-digging, seafood processing, domestic labor, hospitality and hostelry, and the putting-out system of in-house textile production (Meltzoff 1995).

Each occupational activity contributes to the overall production and re-production of the Galician political economy in a closely interwoven pattern. As a result, structural economic changes have a powerful effect on the political economy of daily life and foster support for regionalist responses to these external forces shaping the local social condition. The secondary basis of much Galician agriculture as a supplemental activity and the traditionally small-scale,

Figure 5.3 Traditional Primary Sector Economic Activities in Galicia

personalized social relations characteristic of the fishing industry are reflected by the low level of primary-sector unemployment, which only equals five percent of Galician unemployment (Meltzoff and LiPuma 1986).

The intensification of European Union economic regulation in the 1990s began to generate structural effects among the longstanding traditional Galician modes of primary and secondary sector economic production. These traditional bases of the Galician economy were subjected to rapid transformations that undermined culturally distinct production practices and relations that had been in force in some instances for centuries (García-Ramón et al. 1993). Throughout the 1990s, the rapid transformation of the predominant structure of primary and secondary sector production radically disrupted occupational and household relations among Galician maritime, agro-industrial, and peasant communities as the Galician economy transitioned into a modestly diversified, services-dominated economy. For perspective, from 1970 to 1996 primary sector employment declined from 56.5 percent of the active labor force to only 15.6 percent. The corresponding primary sector contribution to regional gross domestic product declined from 28 percent to only 8.5 percent during the same period (Beiras 1994; IGE 1996). In contrast, tertiary services rapidly doubled throughout the 1980s and 1990s into approximately 50 percent of all employment and almost one-third of the regional GDP (Beiras 1972; IGE 1996).

Perhaps because of these transformations, defense of the traditional Galician primary sector and the spin-off industrial sector became a salient political issue during the 1990s regional elections in disproportion to their size. The primary fishing sector is at the forefront of this Galician political economy. Even by the mid-1990s, the fishing sector generated roughly 57 percent of all Spanish production and directly employed 37,000 workers (IGAPE 1994). Similarly, ancillary and secondary sector maritime industry – shellfishing, aquaculture, seafood canning and packaging, shipbuilding, and such – contributed closer to 80,000 jobs to the Galician and particularly the Pontevedra provincial economy (Ballesteros 2000; Fernández-Macho, Gallastegui, and González 2008). Milk and dairy production, forestry and wood transformation, shipbuilding and steelworks, and a range of mining, textile, and energy production industries constituted the principal primary and secondary mainstays of the Galician economy (IGAPE 1994). Export growth among leading industrial sectors such as transportation manufacturing, granite and slate mining, and steel and metals manufacturing did not counter the negative trade balance of imports over exports in the core agro-industrial sectors such as fishing, energy production, forestry and wood products, and commercial dairy production (IGAPE 1994).

Reconstructing a Galician Space Within the European Union

The EU Maastricht Treaty implementation in 1993 provided Galician political society – with some justification – a focal point for the social unease that permeated everyday life.

The deepening structural and institutional connection between Europe, Spain, and Galicia required the BNG to articulate a political program that would define the territorial organization of political and economic power for the benefit of the Galician economy and culture. The resulting BNG program selectively utilized Galician cultural identity politics to both reflect and feed into growing perceptions of political and economic disempowerment with a regional geographic vision for defending nominally Galician interests. Thus, the BNG political movement came to symbolize and cultivate a core trope of Galician cultural identity bound to the primary and secondary sector employment and a corresponding traditional political economy centered on the peasant-based cultural complex of primary and secondary sector production. The BNG defense of Galician agriculture, dairy, maritime production, and heavy industry prepared the way for the BNG to dominate Galician regionalist representation among regional elections politics throughout the 1990s and into the 2000s.

Accordingly, BNG elections success must take into account the changes to traditional Galician production centered on complex community- and family-based networks of agricultural and dairy production, coastal fishing and shellfish extraction practices, and land tenure regimes little changed from earlier generations (Bauer 1992; García-Ramón et al. 1993; Naylon 1995; Roseman 1996). The late

1980s–early 1990s Galician experience of global political and economic forces reinforced a political-economic 'siege mentality' that opened the door for the consolidation of the BNG as the lone viable heir to Galician regionalist politics. The deindustrialization that eroded the metallurgical and shipbuilding sectors of western Galicia in the late 1970s and 1980s had set the stage for the socio-political transformations to the primary and secondary sectors that accompanied the deepening Galician incorporation into the European-centered political economy during the 1990s period.

Regional politics surrounding the 1993/1997 regional elections solidified around the structural and social importance of Galician primary sector and secondary sector production under threat from European regulation and structural reforms. Major EC/EU policies like the Common Fisheries Policy (CFP) and Common Agricultural Policy (CAP) became discursive rallying points for the BNG political critique and calls to protect the traditional Galician primary and secondary sector production complexes that constituted daily life for many Galician citizens (Meltzoff and LiPuma 1986). European and Spanish political forces represented an externalized threat 'from beyond' to traditional Galician identities rooted in economic class, gender, and place-based identity.

The effects of European Union economic regulation on particularly the 1993/1997 Galician regional elections are visible through review of the European Common Fisheries Policy (CFP) and the territorial reorganization of fisheries governance institutions and practices. From its formal European Community adoption in 1983, the CFP was designed to coordinate biologically and economically sustainable maritime stock management conservation practices through policy instruments like the Total Allowable Catch (TAC) quotas and the 5-year Multi-Annual Guidance Plans (MAGP) focused on maritime fleet reductions and reorganizations (Caballero Miguez et al. 2007). The governance institutions put in place by the CFP generated a complex territorial reorganization of political authority among the EU, Spain, and the Galician government. EU regulatory control over so-called Community waters reinforced the increasingly hierarchical European political-economic framework that removed policy decision-making further away from Galician maritime producers onto higher levels of authority. Notably, the CFP-generated territorial reorganization of fisheries governance ceded Galician coastal waters maritime regulation to the Galician regional government, the *Xunta,* controlled throughout the 1990s by the conservative People's Party (Caballero Miguez et al. 2007).

The complex regulatory geographies that the CFP produced among Galician maritime producers extended throughout the Galician productive complex to transform the social foundations of Galician society centered on small-scale maritime, agricultural and dairy producers and secondary sector maritime processing and shipbuilding industries. Stark reductions in Total Allowable Catch quotas (administered by Spanish officials) and from 1992 mandatory fleet modernization and overcapacity reduction measures had notable direct and indirect effects on the daily lives of Galician peasant producers. As Fernandez-

Macho, Gallastegui, and González (2008) recognize, CFP fisheries policies extended beyond both conservation and economic rationales to have direct effects on the political and social conditions of entire communities and regions effected by European policy decisions.

The political ramifications of EU integration and regulation on the Galician traditional primary and secondary sectors gave sustenance to the BNG political program and drove public support for the BNG in communities and townships hardest hit by EU regulations and perceived governance failures among both Spanish and Galician governments (Nicley 2002). While it is notable that maritime sector employment dipped during the early and mid-1990s, the sector rebounded to a strong economic position by the late 1990s along with substantial European and Spanish financial support for Galician fishing fleet modernization and aquaculture production (Caballero Miguez, Garza Gil, and Varela Lafuente 2007). Nevertheless, the European integration process transformed traditional *genres de vie* that were at the heart of Galician cultural identity. The BNG rise in the 1990s can be seen as a reflection of these societal dislocations that reconstructed the Galician political economy towards closer integration into the European-dominated capitalist political economy. As anthropologist Heidi Kelley (1994, 77) noted:

> Issues of regional identity will continue to be highlighted in coming years, as Galicia struggles to define its role within both the Spanish state and the European Community. As the economy and prestige structure of the community continue to change, this tension [between pride in rural identity and urban cosmopolitanism] will continue to be exacerbated, particularly if regional identity continues to be a locus of resistance to what is perceived as nationally-imposed change.

The complex restructuring of the Galician maritime sector was paralleled in European-led regulation like the Common Agricultural Policy that equally disrupted traditional modes of production and political-economic relations among Galician dairy cattle producers, wood products producers, avicultural and agricultural producers (García-Ramón et al. 1993). For example, increased European market competition during the early 1990s further burdened Galician small-scale dairy producers attempting to comply with Common Agricultural Policy dairy production quotas and overproduction fines levied by the EU against member states and passed down to regional governments like the Galician *Xunta* (El Correo Gallego 1999, 2001).

Taken together, European – Spanish adhesion from 1986 (and particularly the period surrounding the 1993 formation of the European Union) introduced institutional and regulatory reforms that reorganized the territorial governance relations and underlying political economy of Galician society. The 1993 and 1997 regional elections thus reflect the transformed conditions of everyday practice and cultural identity among particular Galician producers and communities.

The political geography of BNG regional elections support can be viewed as a response to the place-specific political and economic dislocations generated by the reterritorialization of political power onto the European Union and concomitant disruption of traditional practices, cultural identities, and social relations in particular Galician communities. Notably, the *minifundia* farm communities hardest hit by European maritime, agricultural, and dairy sector reforms in the early 1990s – 'particularly ... on the coast, where the average farm size is significantly smaller than inland' – also show some of the highest levels of BNG support not only in the 1993 and 1997 regional elections, but throughout the 1985–2005 regional election series (García-Ramón et al. 1993, 8). As anthropologists Edward LiPuma and Sarah Meltzoff (1994, 32) note:

> The problem, and cause for some political discontent, is that the unification (through formalization and standardization) of the institutions of a particular [economic] sector is seen as leading to an internationalization of culture and language and thus a dilution of national and occupational identities ... There is a fundamental link between occupation and Galician identity and ... these are being compromised by EC proposals for restructuring primary industries.

Perhaps foretelling the BNG success later that year, the Spanish daily newspaper *El País* in 1993 published public opinion poll results showing that 49 percent of fishermen believed that the state of the fishing industry had worsened and 54 percent of farmers believed that the state of the dairy industry had worsened, compared to only 7 percent of fishermen and 15 percent of dairy producers who believed their respective industries had improved (Ibañez and Hermida 1993).

Reconstructing the Galician 'Cultural Political Economy': Variation, Selection, Retention

Detailed analysis of the connections between European political economy, Galician regionalist politics, and regional elections returns us to the central question of this chapter: How does the political geography of elections play a role in the reconstruction of cultural political economy? The European project has involved a renegotiation of the political economy among Europe, Spain, Galicia, and particular localities within the Galicia region. European integration policies came together in the historical political moment of the 1990s and in particular places to profoundly change the Galician political economy. The geographic patterns of BNG support reflect the changing Galician landscape of identity politics in the 1990s as communities responded to their everyday experience of capitalist restructuring towards a globalized, European-centered political economy.

However, a cultural political economy perspective on elections takes us beyond the top-down model of a European macro political economy generating Galician regional and place-specific electoral effects. The recursive connection between European integration and BNG support requires that we look more closely at the role of elections in actually reconstructing the European (cultural) political economy from within. The European project established an opportunity for the rise of new political parties responding to transformed political and economic conditions. Yet, the BNG rise was never *a priori* guaranteed in response to the Galician political-economic reforms in the 1990s. The regional elections provided an historical moment for renegotiation of the Galician political economy through competition to define, select, and institutionalize a particular, alternative political-economic vision for Galicia. That is, elections are *constitutive* as an embodied set of practices that reconstruct the cultural political economy within the geographically specific field of identity politics found in particular places.

BNG success in the Galician regional elections over the past 20 years (as well as European, Spanish, and municipal elections) can be viewed as a cultural political economy selection process from among the broader range of Galician political forces and parties. The BNG emerged as the hegemonic voice of Galician regionalist politics throughout the 1990s and our present century through a political 'selection' process in which the alternate political-economic visions (that is, 'imaginaries') offered by regionalist and regionalist forces like the *Coalición Galega* fell aside or were incorporated into existing political discourses. The BNG participation in the Galician electoral process has thus been a platform through which its distinct visions, rhetoric and discursive practices have helped to reconstruct the Galician political economy.

This BNG vision for cultural political-economic reconstruction has centered on the renegotiation of the geographic relationship between Europe, Spain, Galicia, and particular localities within the region. The BNG participation in regional elections – from campaigns, to political demonstrations, to electoral outcomes – has generated greater political space for a Galician regional presence within the overarching framework of the European Union and the Spanish state. The BNG political move from the late 1980s towards a more conciliatory discourse of Galician 'self-determination' within the Spanish state structure and accompanying visions of a strong Galicia within the European Union supports the regionalists' effort to advance the Galician structural position within a European-dominated political economy. The 1994 BNG political rhetoric of 'a strong Galicia in a new Europe' ('*Unha Galiza forte nunha nova Europa*') during EU parliamentary elections reflects the BNG effort to redefine the hierarchical European-Spanish-Galician political economy towards a greater regional voice within the institutional fabric of the European Union (BNG 2010).

Moreover, the Galician regional elections constitute a distinct political moment in the contingent reproduction of the Galician cultural political economy. The regional elections (and particularly the 1993 and 1997 elections) produced complex political-economic visions for Galicia through the contingent articulations of political power among the BNG, mainstream socialist and conservative political forces, and other minor political groups. These contested visions for a new Galician future involved distinct articulations of exactly what elements of society fell within the realm of the political, the economic, and the non-economic. The BNG defense of Galician primary sector and secondary sector producers as cultural standard bearers threatened by the course of European integration and Spanish economic policy found a ready audience among constituents suffering from the economic and social dislocations of the 1990s. The BNG vision for Galicia located traditional production activities and the everyday lives of Galician industrial labor and small-scale agricultural, dairy cattle, and maritime professionals as simultaneously within the realm of the economic and the social – and therefore a legitimate focus for political renegotiation within the European Union framework. The transformed design for the Galician position within Europe and Spain, as articulated notably during regional elections, helped to define and establish the contingent lines of political struggle over cultural and social identity that have become, if fleetingly, the foundations for the production and embeddedness of the Galician political economy within the broader capitalist structure of the European Union. Finally, the BNG role in reconstructing the Galician cultural political economy demonstrates the flexible modalities of capitalist regulation as mediated through culturally contingent political relations. In their call for cultural political economy perspectives, Jessop and Oosterlynck (2008) argue that political economy cannot be reduced to the 'soft sociology' of socially constructed economic discourse unhinged from a broader capitalist structural foundation. The Galician regional elections in the 1990s demonstrate the hybrid conditions of cultural political economy as socially constructed within the bounded possibilities afforded by a culturally mediated capitalist structure. The Galician cultural political economy envisioned by the BNG and BNG supporters (always in relation) centered on reconstructing new economic imaginaries of the Galician position within the European Union. The moderated BNG stance of critical *rapprochement* towards the European project during the 1990s facilitated BNG electoral success and political influence over European integration policy in ways that have continued to yield political results in later years. For example, the European Union inaugurated in 2007 the EU Community Fisheries Control Board headquarters in the Galician industrial port city of Vigo. Such institutional webs that bind Galicia within a common European regulatory framework reflect the contingent political struggles of recent decades to raise the Galician profile politically, economically, and socially within the overarching capitalist structure of the European Union. In this manner, we can see regional elections as a 'sorting out' process in which competing electoral-political forces like the BNG, PSOE, and PP reproduce through electoral processes the broad contours of the European capitalist structure even as the institutional and governance dynamics among Galician intra-community relations, Galician-Spanish relations, and Galician-European Union relations remain in a constant state of evolution towards new forms of culturally grounded political economy.

Conclusions

This chapter explored possible connections between the political geography of elections and cultural political economy. The benefit of cultural political economy is a basic reminder that the capitalist socio-economic structure is a social product that takes shape through a geographically contingent mix of economic and identity politics. In other words, capitalist political economy is always in a state of being reproduced and that reproduction process opens the door for the transformation of political-economic structure through political struggle. Following Jones (2008), I have suggested that we can approach elections as a particular political moment in the contingent reproduction of a culturally mediated capitalist political economy.

The Galician regional elections and the rise of the Galician Nationalist Bloc (BNG) provided a strong case of capitalist political economy as a shaping factor in electoral outcomes. At the same time, European political and economic integration afforded an opportunity window for the negotiated reconstruction of the European-centered political economy through the political struggles among European, state-range, and regionalist political forces to redefine and consolidate a distinct political-economic vision for the future. The Galician regional elections highlighted the complex geography of these reconstruction processes. Geographic concepts of power, place, and scale are central to the ongoing reconstruction of European-centered cultural political economy, from the struggles over territorial governance between the EU, Spain, and Galicia to the place-specific political struggles over political, economic, and cultural identity that generated bases of support for the BNG program to rebuild the Galician political-economic position within a dominant European capitalist system.

Approaching the political geography of elections as a point of entry for research on the production and changing characteristics of capitalist political economy gives political geography an opportunity to bring together research on macro-political and macro-economic forces and the contingent, grounded political practices that go into reconstructing capitalist political economy on a daily basis. The connections between elections and cultural political economy remain open to significant refinement, clarification, and modification – this chapter offers only an initial consideration of the possible avenues that political geographers might explore. However, the potential benefits for political geography rest in the opportunity to reinvigorate electoral studies on the leading edge of contemporary social theory on the intersections between capitalist regulation and post-structural thought.

References

Agnew, J. 1997. The dramaturgy of horizons: Geographical scale in the 'Reconstruction of Italy' by the new Italian political parties, 1992–95. *Political Geography* 16 (2): 99–121.

Agnew, J. 2000. From the political economy of regions to a regional political economy. *Progress in Human Geography* 24: 101–10.

Aizpeolea, L. 1997. El gobierno atribuye el alza del nacionalismo gallego a su componente de izquierdas. In *El País* print edition. Published October 21, 1997. Available at: http://www.elpais.com/articulo/espana/espana/galicia/espana/bloque_nacionalista_galego_BNG/partido_popular_/PP/partido_socialista_obrero_espanol_/PSOE/elecciones_autonomicas_gallegas_2001_/21-10-2001/elpepiesp/19971021elpepinac_4/Tes [last accessed: October 27, 2010].

Ballesteros, M.A. 2000. Technology and employment, new ways of organising production and labor. Modernisation of traditional economic industries: The fishing in Galicia (Spain). The public administration as advocacy actor. Paper presented at The Ionian Conference: Facing the Challenges of the New Millennium, at Corfu, Greece.

Barreiro Rivas, X.L. 2003. Da UPG ao BNG: O proceso organizativo do nacionalismo galego. In X.M. Rivera Otero (ed.), *Os Partidos Politiicos en Galicia*. Vigo: Xerais de Galicia, 99–262.

Bauer, R.L. 1992. Changing representations of place, community, and character in the Spanish Sierra del Caurel. *American Ethnologist* 19 (3): 571–88.

Baylina, M., and M.D. García-Ramón. 1998. Homeworking in rural Spain: A gender approach. *European Urban and Regional Studies* 5 (1): 55–64.

Beiras, X. 1994. *O Atraso Económico de Galicia*. Second edition. Santiago de Compostela: Galaxia.

BNG. 2010. *A Historia do BNG* [BNG Official Website]. Available at: http://www.bng-aliza.org/opencms/opencms/BNG/global/contidos/textos/historia/textos/texto_0001.html [last accessed: October 20, 2010].

Botella, J. 1989. The Spanish 'new' regions: Territorial and political pluralism. *International Political Science Review* 10 (3): 263–71.

Caballero Miguez, G., M.D. Garza Gil, and M.M. Varela Lafuente. 2007. Institutions, governance and fishing policy: Theoretical foundations and the analysis of the Galician case. Paper read at European Association for Evolutionary Political Economy (EAEPE), at Porto, Portugal.

Casal, C., and L. Ventoso. 1994. *Un Solo de Gaita*. Santiago de Compostela: Biblioteca Gallega.

Clayton, T. 2002. Politics and nationalism in Scotland: A Clydeside case study of identity construction. *Political Geography* 21: 813–43.

El Correo Gallego. 1999. Os criadores avícolas, contra as propostas de UE. *El Correo Gallego (Digital edition)*, 15 June 1999.

El Correo Gallego. 2001. Milleiros de gandeiros reclaman en Santiago axudas directas para o sector. *El Correo Gallego (Digital edition)*, April 8, 2001.

Fernández-Macho, J., C. Gallastegui, and P. González. 2008. Economic impacts of TAC regulation: A supply-driven SAM approach. *Fisheries Research* 90 (1–3): 225–34.

Flint, C. 1996. Whither the individual, whither the context? *Political Geography* 15 (2): 147–51.

Flint, C. 2001. A timespace for electoral geography: Economic restructuring, political agency and the rise of the Nazi Party. *Political Geography* 20: 301–29.

García-Ramón, M.D., M. Vilarino, M. Baylina, and G. Canoves. 1993. Farm women, gender relations and household strategies on the coast of Galicia. *Geoforum* 24 (1): 5–17.

Gibson-Graham, J. 1996. *The End of Capitalism (As We Knew It): A Feminist Critique of Political Economy*. Cambridge, MA: Blackwell.

Ibañez, J.G., and X. Hermida. 1993. Una ligera mayoría cree que en Galicia se vive mejor. *El País*, October 11, 1993, 17.

IGAPE. 1994. *Galicia: Good Business Opportunities*. Edited by I.G.d.P. Económica. Santiago de Compostela: Xunta de Galicia.

IGE. 1996. Padrón de Habitantes de 1996: Instituto Galego de Estatística.

Jessop, B., and S. Oosterlynck. 2008. Cultural political economy: On making the cultural turn without falling into soft economic sociology. *Geoforum* 39: 1155–69.

Jessop, B., and N.-L. Sum. 2001. Pre-disciplinary and post-disciplinary perspectives in political economy. *New Political Economy* 6 (1): 89–101.

Jessop, B., and N.-L. Sum. 2007. *Beyond the Regulation Approach: Putting Capitalist Economies in Their Place*. Cheltenham, UK and Northhampton, MA: Edward Elgar.

Jones, M. 2008. Recovering a sense of political economy. *Political Geography* 27: 377–99.

Keating, M. 2001. Rethinking the region: Culture, institutions and economic development in Catalonia and Galicia. *European Urban and Regional Studies* 8 (3): 217–34.

Kelley, H. 1994. The myth of matriarchy: Symbols of womanhood in Galician regional identity. *Anthropological Quarterly* 67 (2): 71–80.

LiPuma, E., and S.K. Meltzoff. 1994. Economic mediation and the power of associations: Toward a concept of encompassment. *American Anthropologist* 96 (1): 31–51.

MacLeod, G. 2001. New regionalism reconsidered: Globalization and the remaking of political economic space. *International Journal of Urban and Regional Research* 25 (4): 804–29.

Marden, P. 1997. Geographies of dissent: Globalization, identity and the nation. *Political Geography* 16 (1): 37–64.

Márquez Cruz, G. 2004. Las elecciones locales y provinciales en Galicia (1979–2004). 1. Comportamiento electoral y continuidad/renovación de las autoridades locales. *Revista de Investigaciones Políticas y Sociológicas* 3 (2): 111–33.

Meltzoff, S.K. 1995. Marisquadoras of the shellfish revolution: The rise of women in co-management on Illa de Arousa, Galicia. *Journal of Political Ecology* 2: 20–38.

Meltzoff, S.K., and E. LiPuma. 1986. The troubled seas of Spanish fishermen: Marine policy and the economy of change. *American Ethnologist* 13 (4): 681–99.

Naylon, J. 1995. On the periphery: A view from rural Galicia. *Leeds Papers on Spain in Europe* n.d.: 39–79.

Nicley, E. 2002. Somos Galegos!: Regionalism, political economy, and place context in Galicia, Spain. Master's Thesis, Department of Geography, Florida State University, Tallahassee, FL, USA.

Parlamento de Galicia. 2010. *Resultados electorales*. Published by the Parliament of Galicia. Available at: http://www.es.parlamentodegalicia.es/sites/ParlamentoGalicia/ContenidoGal/pargal_II_ResultadosElectorais.aspx [last accessed: October 10, 2010].

Portas, P. 2010. The myth of unity and the BNG: Storytelling, nationhood and power. In *Political Studies Association*. Edinburgh: Unpublished manuscript.

Rivas, M. 1997. ¿Por qué sube Beiras? *El País* print edition. Published October 13, 1997. Available at: http://www.elpais.com/articulo/espana/espana/galicia/espana/Bloque_Nacionalista_Galego_/BNG/elecciones_autonomicas_Gallegas_1997_/19-10-1997/sube/Beiras/elpepiesp/19971013elpepinac_11/Tes?print=1 [last accessed: October 29, 2010].

Roseman, S. 1996. 'How we built the road': The politics of memory in rural Galicia. *American Ethnologist* 23 (4): 836–60.

Roseman, S. 1999. ¿Quen manda? (Who's in charge?): Household authority politics in rural Galicia. *Anthropologica* 41 (2): 117–32.

Sayer, A. 2001. For a critical cultural political economy. *Antipode* 33 (4): 637–708.

Sharif, G. 2006. *Galicia: A Concise History*. Cardiff: University of Wales Press.

Shin, M. 2001. The politicization of place in Italy. *Political Geography* 20 (3): 331–52.

Vilas Nogueira, J. 1992. *Las elecciones en Galicia (1976–1991)*. Barcelona: Institut de Ciències Polítiques i Socials.

Chapter 6

Place, Identity, and Taiwan's Presidential Elections, 1996–2008

Daniel McGowin

On 23 March 1996, the Taiwanese people celebrated a monumental day in the history of the Republic of China (ROC). On that day, *Kuomintang* (KMT) nominee Lee Teng-hui beat out two opponents to remain the president of the ROC. This election was significant not because Lee was re-elected, but it marked the first time that the Taiwanese people were able to directly elect the leader of the country.

Previous to the 1996 presidential election, the now defunct National Assembly held the responsibility of selecting the president of the ROC. The National Assembly, established in 1948 in accordance to the constitution of the ROC, only elected into office two presidents during its existence – Lee Teng-hui in 1988 and Chiang Ching-kuo in 1978.[1] When a new National Assembly was established in 1991, the decision to abandon indirect presidential elections was made and the process towards the country's first direct presidential elections commenced. This culminated in the historic re-election of Lee in 1996.

Since this election, Taiwan has held direct presidential elections every four years, with the most recent occurring in 2008. During the past 12 years, Taiwan experienced a shift in political parties, with the opposition Democratic People's Party (DPP) winning both the 2000 and 2004 presidential elections. By the time a KMT candidate returned to the presidency in 2008, a clear geographic pattern had emerged among the Taiwanese electorate. While the 1996 elections in Taiwan were a landslide victory for the KMT in terms of counties and municipalities won, the 2000 elections experienced a dramatic change in the electoral geographic landscape. Due to a split in the KMT, the opposition DPP was able to garner a plurality of votes and its nominee Chen Shui-bian won the presidency. With this election, a political cleavage emerged in the southwestern part of the island. Chen succeeded in winning re-election in 2004, where the distinct voting pattern from 2000 held true. With the 2008 victory by KMT nominee Ma Ying-jeou, there was a shift back to the old guard. However, the pattern that emerged in 2000 was once again reinforced in the 2008 presidential elections.

---

1 Yen Chia-ken, who succeeded Chiang Kai-shek, was vice president at the time of Chiang's death and therefore ascended to president without the National Assembly electing him to the position.

Because political parties in Taiwan tend to carry a particular position on Taiwan's international status, the emergence of such patterns are likely linked to locally-held opinions on independence, as well as questions of identity and regional rivalries. The purpose of this chapter is to explore the voting patterns of the four presidential elections in Taiwan and how these patterns represent the ongoing debate over Taiwan's international status. The chapter seeks to further the work done by electoral geographers linking votes and elections with debates over identity, and add to the limited number of studies of Asian elections with the English-language electoral geography literature (see Lee and Brunn [1996] for one example from South Korea).

The chapter begins with a history of political parties in Taiwan, followed by discussion of the intersections of identity politics and political parties. The chapter concludes with an analysis of the 1996, 2000, 2004 and 2008 presidential elections, and suggests how place and identity are crucial concepts in understanding the geographic patterns of these votes.

Political Parties in Taiwan

The Origins of the Kuomintang

Since its founding, the Republic of China has been dominated by the *Kuomintang* (KMT), be it in the capacity of the presidency, in the legislature, or in the local offices. This dominance was so prevalent that the KMT was synonymous with 'Nationalist China,' a term used to differentiate the ROC from the People's Republic of China (Wright 1955).

The roots of the KMT date back to the Qing Dynasty (Booth 1999). The original purpose of the KMT was revolutionary in nature; 'a revolution to establish the Republic' (KMT website 2009). Once the Republic of China was established and the KMT gained power over the mainland in 1912, the ideology of the party shifted from one of revolution to one of state-building. However, the wars with the Communist and with Japan made it difficult for the KMT to establish itself internationally. Despite Western support, the Nationalist suffered numerous losses to the Communist and eventually fled to the island of Taiwan. It was here that a new KMT ideology developed, one that continues to influence the party even today.

However, it is important to note the transformation of the KMT's ideology did begin while the party and the ROC controlled the mainland. The switch from a revolutionary ideology to a restoration ideology was based on the need to legitimate its claim as the leader of China. In this sense, the KMT felt it best to project itself as a continuation of Chinese history rather than a completely new beginning. Party officials, including Chiang Kai-shek, also sought to promote the Confucian traditions that were prevalent, albeit intermittently, throughout China's long history (Wright 1955). Yet, the KMT's shift was also influenced by interaction with the Soviet Union, as in 1924 Sun Yat-sen adopted Leninist elements from the Soviets, even going as far as to using the Soviet Communist Party as a model for reorganization (Dickson 1996). This Leninist influence played a significant role in the reorganization of the KMT once it fled to Taiwan.

The Reorganization of the KMT and Its Current Philosophies.

The goal of the KMT/ROC was never to remain on the island of Taiwan, but to reorganize and retake the mainland from the Communist. But coexisting on the island with the native Taiwanese would prove to be a difficult task.

The term 'native Taiwanese' refers to those people who arrived on the island previous to the ROC gaining possession of the island in 1945. Bedford and Hwang (2006, 5) note that:

> 85 percent of Taiwan's current residents have ancestors who emigrated from China three- to four-hundred years ago ... The remaining 13 percent of nonindigenous residents arrived in Taiwan or are descended from those who arrived in Taiwan between 1945, when Japan ended its colonization, and 1949 when Chiang Kai-shek and his armies and followers arrived.

This latter group is referred to as *waisheng*, which references their 'foreign' nature to Taiwan. This distinction, while not necessarily visible, is an important ethnic difference that plays out in present-day Taiwanese politics. Because the KMT were viewed as outsiders and were the minority on the island, they needed a way to consolidate power.

The KMT did little to enamor themselves with the Taiwanese in the immediate years following the end of the Pacific War. For all the negative actions that the Japanese took during the Pacific War, it is also clear that they modernized and invested in Taiwan (Cumings 1984). In fact, by the time the Japanese were forced out, Taiwan had actually advanced past China. This notion was reinforced by the actions of the Chinese army that arrived on the island. Due in part to receiving low pay, the soldiers plundered the local community. Furthermore, the KMT 'dismantled Taiwanese infrastructure including telephone wires, pipes, metal roofing, fire hydrants, and railroad switches and shipped everything back to the mainland to support the effort there, asserting that Taiwan's citizens owed a debt to the KMT for liberating them from the Japanese' (Bedford and Hwang 2006, 6). As crime rates rose and the KMT took control of major industries, a feeling of resentment began to fester. The local feeling at the time is summed up best by a local saying quoted by Bedford and Hwang (ibid.) 'that the dogs (who could at least protect your property) had been chased away, but the pigs (who only make a mess) had come.'

Taiwanese resentment culminated in what is known as the 228 incident. Beginning on 27 February 1947, protests began in Taipei and spread throughout the country. The response was brutal as the Republic of China's army, backed by the KMT, violently suppressed the uprising (Wu 1995; Cooper 1998; Huat 2001). This anti-government revolt not only galvanized Taiwanese resentment towards the KMT, but it also solidified a mainland identity to the KMT. The 228 incident 'resulted in considerable bloodshed after soldiers were called from the mainland to put it down, and in its aftermath there was deep antipathy on the part of the Taiwanese toward Mainland China, the KMT, and the government' (Cooper 1998, 15). The subsequent suppression of the majority Taiwanese by the KMT left an impact that continues to define the political landscape in Taiwan. As Huat (2001, 135) notes:

> local Taiwanese suffered a regime of 'white terror' under the KMT's security apparatus, identified with the colonizing mainlanders, who constituted no more than 15 percent of the population. The terms 'mainlanders' and 'locals' thus became an ideological/identity/political divide providing one of the central discursive resources for the organization of opposition to the KMT and preventing the formation of a Taiwan 'nation' so long as the 'mainlanders' are in power.

Thus, the KMT was ascribed with the mainlander identity, regardless of the ethnicity of its members. Additionally, any party that came to oppose the KMT could potentially be ascribed with a Taiwanese identity.

After the ROC leadership fled the mainland for Taiwan, the KMT struggled to find its place on the island. While the KMT initially attempted to undercut identity politics by sinicizing the Taiwanese (Huat 2001), the party soon realized that it needed to adapt to its new political environment. As Dickson (1995, 49) notes that 'the main goals and tasks of the KMT changed: Rather than concentrating on plans to retake the mainland, the party devoted more of its energy to issues of immediate concern to Taiwan and its own reputation.' One way to accomplish this was to Taiwanize the government, as well as the party itself, by allowing Taiwanese to join its ranks. In order to do this, the KMT utilized local elections to integrate Taiwanese into the political process. However, the central government remained in the hands of the KMT. This integration process created the façade of an all-inclusive government while maintaining the legitimacy of the ROC state (Rigger 2001). A similar tactic was repeated in the 1970s as the KMT government attempted 'to maintain the legitimacy of its power' (Bedford and Hwang 2006, 7). In this case, Taiwanese elites were allowed to join the party.

Not surprisingly, this shift in attitude coincided with an international shift in recognition. KMT leadership recognized that reclaiming their stake to the mainland would have to be put off and a new focus would need to be shift to simply maintaining some form of legitimacy. There was some relaxing of the KMT's political control, but martial law, for example, remained in effect until the mid-1980s. Taiwanization of the party placed Taiwanese members into local offices and helped integrate Taiwanese, including future president Lee Teng-hui, into the central government (Dickson 1995). While change was slow, by 1985 the influx of Taiwanese members of the KMT had made a profound impact. Using data gathered by Yang (1985), Dickson (54–5) notes that 'the proportion of Taiwanese among district cadres rose to 56.6 percent in 1975 and to 73.3 percent in 1985. At the county level, the proportion of Taiwanese increased from 34.5 percent to 53.9 percent between 1975 and 1985.' While some of this could be explained by the lack of immigration from the mainland – and therefore fewer mainlanders from which to choose – this pattern marks a new trend within the KMT; one implemented by Chiang Ching-kuo that differed from his father Chiang Kai-shek, who preferred mainlanders to Taiwanese. And even though the numbers of Taiwanese in the provincial cadres were not as high, the impact of locally visible Taiwanese KMT members was important. 'In local offices where society came into direct contact with the party, the increased visibility of Taiwanese in party posts created at least the appearance of greater responsiveness to local wants and needs, a basic goal of the Taiwanization strategy' (Dickson 1996, 55).

The reign of Chiang Ching-kuo also marked a departure from the tight-fisted control demonstrated by his late father. The younger Chiang began to loosen up restrictions on opposition parties, something that was forbidden during the era of martial law. In 1986, the opposition Democratic Progressive Party (DPP) was established and Chiang took 'a tolerant position' (KMT website 2010). One year later, martial law was officially lifted on the island. These steps not only helped establish an opposition to the KMT, but it also provided a much louder voice for the Taiwanese people; a voice that signaled the beginning of ethnic politics.

From Dangwai to Opposition: The Rise of the DPP

One way the KMT kept Taiwanese from directly electing the president, or even establishing political parties, was the notion that the Republic of China was not confined solely to the island of Taiwan. Indeed the KMT-controlled ROC stressed the fact that they represented the mainland. Therefore, to allow the people of Taiwan to be the only ones selecting the leader for all of China would disenfranchise their fellow countrymen and women on the mainland. While this line of reasoning worked in the 1950s and 1960s, the changing political climate of the 1970s eroded this façade and led the KMT to reconsider its position.

Although the KMT dominated politics on the island, there were a few independent candidates who were able to break through and reach the upper echelon of the Taiwanese political hierarchy. One of the first was Kao Yu-shu (Henry Kao), who was able to defeat KMT opponents to become mayor of Taipei in 1954 and then again in 1964. Kang Ning-hsiang was able to work his way into the Legislative Yuan by 1972. Since he was not a member of the *Kuomintang*, he took up the moniker of a *Dangwai* candidate, or an 'outside the party' candidate. This denotation, once held in a negative light and seen as a threat to the KMT, is an important step towards creating an organized opposition to the KMT. Although the *Dangwai* had its share of factionalism (Rigger 2001), the rise of *Dangwai* candidates in the 1970s helped stoke the fire of opposition in Taiwan. It is from this movement that the DPP gained its initial momentum.

The wave of democratization and the loosening of the authoritarian system on the island began in the early 1970s. Triggered by the supplementary elections to replace members of the Legislative Yuan who either retired or passed away, non-KMT politicians began to break into the political system. However, it would be a mistake to assume that the opposition held any power or that it was a solid cohesive unit. There were some attempts to organize the opposition movement, most notably the *Dangwai* and its accompanying political publication *Formosa*. The magazine set up offices throughout the island and helped create a network in which to build opposition. The threat posed by the *Dangwai* via *Formosa* was very real in the eyes of the KMT leadership. The rising opposition and the KMT's fear culminated in a clash in 1979 in the southern city of Kaohsiung. This incident, known as the Kaohsiung Incident or the Meilidao Incident, occurred on 10 December 1979 as opposition leaders held protests against the KMT-led government. Police, both civilian and military, responded with violence and several opposition leaders, including those that did not participate in the protest were jailed (Rigger 2001; Bedford and Hwang 2006; Wu 1995). While some of the leaders were removed from the political process, the fallout of the Kaohsiung Incident led to a rise of new opposition leaders, including Chen Shui-bian, who was an attorney for the jailed dissidents (Rigger 2001).

The Kaohsiung Incident marked a transition into a comparatively more relaxed political climate. Ching-kuo allowed for some publications to continue printing dissenting voices. One such publication was *The Eighties*, headed by moderate opposition leader Kang Ning-hsiang. Kang represented a more tolerable opposition in comparison to the *Formosa* group. However, despite the arrests and depletion of activist opposition leaders, there were still factions in the *Dangwai* movement that challenged the moderate position, namely the defense lawyers and a new group of radicals (Rigger 2001; Wu 1995). While some of the defense attorneys and even family members of jailed dissidents won seats in the 1980 elections, the *Dangwai* movement factionalized even more. Radical magazines such as *Deep Plow* accused the moderate Kang of having close ties with the KMT. When the two sides attempted to work together in 1983, disagreement arose about how the selection process should be conducted; moderates preferred a top-down approach while the activists favored a bottom-up method. By 1984, there were at least four factions: 'the Writers and Editors Alliance, the Dangwai Public Policy Research Association (DPPRA), the Progress Faction … and the imprisoned Formosa Faction' (Rigger 2001, 23). The radical Writers and Editors Alliance would soon establish the *New Tide Magazine*, which continued to be the voice of the radical faction until 2006 when DPP factions were abolished (DPP website 2010; Rigger 2001).

Despite factionalism, the *Dangwai* collectively pursued the establishment of a political party. Under martial law, political parties other than the KMT were illegal. Nevertheless, the opposition recognized the importance of having a political party. In 1986, Kang established a DPPRA branch in Taipei. After that time, the opposition spent the summer carefully planning their move. On 28 September, the opposition arrived at a consensus and established the Democratic Progressive Party. 'The party represented a coalition of all camps within the Dangwai: the Mainstream Faction (or Kang Group), New Tide (formerly the Writers and Editors Alliance), and Lin Cheng-chieh's Progress Faction. The Formosa Faction was absent at the founding only because the leading Formosa activists were still behind bars' (Rigger 2001, 25). It is important to note that the party was founded despite the fact that such parties were illegal. However, Chiang Ching-kuo, who had advocated for reform, turned a blind eye to the DPP and held back conservative calls for prosecution (Wu 1995).

The DPP wasted little time breaking into the upper tier of the Taiwanese political system, making headway in 1986, 1989 and 1992. By March 2000, just shy of 14 years after its founding, the DPP claimed the presidency in Taiwan's second direct presidential election.

What does the 2000 victory by Chen Shui-bian represent for Taiwan? Or, more importantly, how does this victory reflect the identity debate in Taiwan? To answer this, it is imperative to understand the identities of the political parties of Taiwan, namely the KMT and DPP.

Identity Politics and Political Parties in Taiwan

As noted above, the KMT carries with it the mainlander identity. This denotation is due in large part because of the migration of the party from the mainland to Taiwan. However, the connection runs deeper than a simple transfer of location from Nanjing to Taipei. Under Chiang Ching-kuo, the KMT went to great lengths to Taiwanize the party. However, events such as the 228 Incident and the Meilidao Incident injured the party's reputation. With the KMT resembling more of a colonizer rather than fellow citizens during the early years, the KMT was stained with the outsider stigma. Regardless of the steps taken by Ching-kuo, Lee Teng-hui and other party leaders, the KMT continues to wear the mainland label.

The DPP, by way of comparison, carries with it the party of the Taiwanese. The 228 incident, which created a great rift between the Taiwanese and the mainlanders, remains an important symbol of the opposition movement. That incident came to represent the resistance against the mainlanders and, by extension, the KMT (Hsieh 2008). Through this connection to the 228 Incident, as well as the Meilidao Incident, the birth of the DPP was in direct opposition to mainlanders as a whole, and the KMT in particular. Through these means, the DPP is identified as 'Taiwanese.'

However, to breakdown the KMT and DPP solely on the basis of ethnic identity is erroneous. While the parties may have carried a strict ethnic identity early on, especially the KMT, that is no longer the case. As noted above, through the efforts of Chiang Ching-kuo, the KMT has become more diversified in its ethnic make-up. According to Huang (1996, 115), the ethnic composition of the KMT shifted from 60.6 percent mainlanders in 1969 to only 30.8 percent in 1992. Additionally, the composition of the KMT Central Committee grew from 3.1 percent Taiwanese in 1952 to 53.3 percent Taiwanese in 1993 (118). This is also reflected in the electorate, as Taiwanese and Mainlanders were equally likely to support the KMT (Rigger 2001, 164). This trend indicates that the KMT's composition has lost its mainlander identity, even if it continues to be painted as a mainlander party. The reason for this has more to do with support for the DPP. As Rigger notes, 'Taiwanese were far more likely to support the DPP than were Mainlanders.' Thus, Rigger concludes, the characterization of the DPP as the party of the Taiwanese remains an accurate one. If the KMT is to continue to carry with it the identity of the mainland, it is only because of its historical antecedents and by way of contrast to the DPP rather than anything based on reality.

However, the primary difference between the DPP and KMT rests on the issue of national identity rather than ethnic identity. Hsieh (2008, 12–3) notes that 'as Taiwan became more and more democratic, Taiwanese, as the majority population, gradually got the upper hand, and the salience of ethnicity in Taiwanese politics declined. In its stead, the issue of national identity became more and more prominent in Taiwan's politics.' It is important to note what is meant by 'national identity.' In this case, the sides fall either for independence (from China), the status quo, or reunification (with China). Political parties that are considered to be part of the pan-Blue coalition, namely the KMT, People First Party (PFP) and the New Party (NP) tend to support the status quo with a slight lean towards reunification. The pan-Green coalition tends to support independence and is led by the DPP, the Taiwan Solidarity Union (TSU) and the appropriately named Taiwan Independence Party (TAIP). It is worth noting here that support for independence or for status quo/unification does not necessarily fall along ethnic lines. The issue of national identity has existed since the KMT arrived on the island. While this issue coexisted along with the ethnic divide, the Taiwanization of the KMT and the rise to power of Taiwanese has lessened the role of ethnicity and given way to the effect of national identity. Nevertheless, there are some mainlanders who favor independence and some native Taiwanese who support the status quo. The issue of national identity could not easily fall along ethnic lines because, as Bedford and Hwang (2008, 11) note, 'with native Taiwanese constituting the vast majority of the population, if the independence-unification issue were decided along purely ethnic lines, the pan-blues would have no hope of getting elected.'

That the national identity issue is the main division between the two major parties, and more generally the two coalitions, is portrayed in a table provided by Rigger (2001, 165). In a post-election survey after the 1995 legislative election, 41 percent of Taiwanese and 44.7 percent of mainlanders were likely to support the KMT (21.9 percent and 5.4 percent, respectively, for DPP support). In terms of self-identification, 34.5 percent of those who consider themselves Taiwanese supported the KMT while 51 percent of self-identified Chinese sided with the KMT. In general, Taiwanese and mainlanders were equally likely to support the Kuomintang. However, when it came to the question of national identity, 49.5 percent of unification supporters and 45.7 percent of those that wanted to maintain the status quo preferred the KMT. In contrast, only 20.2 percent of supporters for independence were likely to back the KMT. Conversely, 52.8 percent of voters who backed independence preferred the DPP, with supporters of unification (10.7 percent) and status quo (15.7 percent) not likely to favor the DPP.

What these numbers suggests is that the division between the two main Taiwanese political parties is along national identity issues rather than ethnic issues. However, it would be erroneous to suggest that this is clear cut, that the national division is the only difference between the parties, or that ethnic identity plays no role. The characteristics of the KMT are more muddied than that of the DPP. In fact, those that support independence who still backed the pro-unification party was double the percentage of those unification supporters who still preferred the pro-independence DPP. As Rigger concludes, 'party identification is not a simple matter of ethnic identity or policy preference.' However, it is clear that the DPP elicits a certain type of voter – ethnic Taiwanese; pro-independence – than does the KMT. Thus, the supporters of the DPP are more defined by their ethnic composition and policy preference than the KMT. Ethnic identity is still important with regards to the DPP given the parties pro-independence lean – a policy that resonates well with many Taiwanese. With the number of self-identified Taiwanese tripling since the 1990s (Bedford and Hwang 2006), ethnic identity continues to be important for the DPP.

What can be stated here is that because of how the supporters of the DPP are defined, supporters of the KMT are 'othered' as being mainlanders and pro-unification, even if it is not entirely accurate. Yet, the ethnic divide has become less of an indicator of support while the national identity issue continues to provide a clearer understanding for one's support for a given party. Certainly ethnic identity is an indicator of support for independence (Rigger 2001; Hwang and Bedford 2006; Hsieh 2008), but the national identity issue still remains the better determinate for party support in Taiwan. And, given the platforms of the two parties, the national identity issue is still a central division to the parties itself even if the electorate may not be as clearly defined, at least with regards to the Kuomintang.

Thus, how does geography aid in better understanding this division? Does place prove to be a clearer determinate for party support than ethnic identity or policy preference? In turn, how does place reflect the debate over national identity; in other words how does geography explain the discourse over what it means to be 'Taiwanese'? As a way to illustrate this, the remainder of this chapter will examine direct presidential elections since 1996. However, before doing so it is necessary to briefly explain the process of selecting Taiwan's president prior to 1996.

The Democratic Transition in Taiwan

As noted above, the National Assembly, the constitution-amending body of the Taiwanese government, elected the president for the Republic of China prior to

1996. But, elections at all levels have been limited to some degree by the KMT-controlled government. The freeze on elections for legislative members as well as the ban on political parties created a one-party system that ensured the KMT remained in power while it attempted to reorganize and plan its return to the mainland. With the shift in attitudes in the international community, the KMT began to soften its stance and slowly allowed for a more open and democratic system to take effect.

Tien (1996) notes that the transition from an authoritarian system of government to modern democracy took place in three phases. The first stage, from the 1950s until 1971, witnessed elections at a local level. The first local elections took place in 1952, allowing Taiwanese to vote for town chiefs, county and city councilpersons, and city mayors. Direct elections were not allowed above the provincial level. During this time non-KMT candidates were able to receive as much as 35 percent of the vote. However, because of the implications of martial law, these non-KMT candidates could not organize as a party. In fact, the KMT banned candidates from engaging in political activities outside of their voting districts.

The second period (1972–85) was marked by the decline in international recognition of the Republic of China. As noted above, as a means of maintaining some semblance of political legitimacy the KMT began to Taiwanize its party and by extension Taiwanize the political hierarchy. The KMT also began to allow an expansion of the election process in Taiwan. There were supplementary elections for the National and Legislative assemblies in 1969 and 1972, then more regularly beginning in 1980. The *dangwai* were allowed to participate as a quasi-party and began making their way into the various assemblies. Tien attributes this expansion, in part, to the rising economic condition in Taiwan. Economic growth in Taiwan led to more opportunities for higher education attainment, greater personal wealth, a larger middle-class, and a more diverse media industry. 'These developments provided not merely the requisite conditions for democratic change but also the additional driving forces for liberalizing the political system.'[10] Additionally, more Taiwan-born political activists meant fewer connections to the mainland and a swing in momentum in favor of opposition. This trend set the stage for the third phase.

With the birth of the DPP in 1986, the third phase opened the door for elections under a competitive party system. However, it was not until after the death of Chiang Ching-kuo in 1988 that political parties were legalized. Under the leadership of Lee Teng-hui, more steps were taken to liberalize the political system of Taiwan. The most important here is the move to direct elections for president. Lee worked with both KMT and DPP leaders to hatch out a plan to allow the voters of Taiwan to decide on their president. In 1992, the Legislative Assembly passed an amendment that would abolish the previous method for electing the president. However, the amendment did not immediately decide on a new method, choosing instead to wait until one year before the next election, which was in 1996 (Lu 2002). The amendment to the 1947 constitution, which was enacted in 1994, also set the length of terms to four years each with a maximum of two consecutive terms (Lu 2002; Myers et al. 2002).

The effects of the democratization process of Taiwan can be seen in the early open elections of the Legislative Assembly. The *Dangwai*/DPP saw its representation in the Legislative Yuan jump from 13 percent in 1980 to 33.2 percent in 1995 before sliding down slightly to 29.6 percent in 1998 (Tien 1996, 16; Diamond 2001, 53). Additionally, Diamond notes that by 1995, the KMT's representation in the Legislative Yuan dipped below 50 percent. Similarly, the KMT's representation among county magistrates and city mayors fell below 50 percent in 1993 (Tien 1996, 17). This pattern culminated in the 2000 election of Chen Shui-bian as president, giving the DPP a plurality of the votes. Four years later, Chen secured just over 50 percent of the votes for the DPP.

Since this election, Taiwan has held direct presidential elections every four years, with the most recent occurring in 2008. During the past 12 years, Taiwan experienced a shift in power, with the opposition Democratic People's Party (DPP) winning both the 2000 and 2004 presidential elections. By the time a KMT candidate returned to the presidency in 2008, a clear geographic pattern had emerged among the Taiwanese electorate. And because political parties in Taiwan tend to carry a particular position on Taiwan's international status, the emergence of such patterns are likely linked to locally-held opinions on independence, as well as questions of identity and regional rivalries.

Four Presidential Elections in Taiwan: 1996–2008

The presidential election in 1996 witnessed an overwhelming victory for the KMT, with their candidate Lee Teng-hui winning 24 of the 25 districts. Independent candidate Lin Yang-gang, who finished third in the voting, captured his home county of Nantou in central Taiwan. Lee garnered a majority of the votes in 18 of Taiwan's 25 districts and at least 40 percent of the votes in five other districts. The only two districts where he received less than 40 percent of the votes were in the aforementioned Nantou County (31.5 percent) and Taipei City in the north (38.9 percent). Thus, the 1996 election pattern represents a fairly solid voting bloc in Taiwan as the KMT was able to secure victory throughout the country.

While the 2000 presidential election involved five candidates, the competition was primarily between three individuals – Lien Chan of the KMT, Chen Shui-bian of the DPP, and James Soong of the newly formed People First Party (PFP). With 39.3 percent of the vote, Chen won the election and for the first time it ROC history the country was led by someone not of the KMT party. Not only did the KMT lose, Lien did not win a single electoral district. While Chen won the popular vote, Soong actually won more districts – 15 to Chen's ten. Here, a geographic pattern begins to emerge as the DPP won districts in the south and west as well as Yilan County in the north, and Soong won a contiguous string of districts from the west to the north, including the three western island districts – Kinmen County, Lienchiang County and Penghu County. It should be noted that Chen only gained a majority of votes in one district, his home county of Tainan, while Soong managed to receive a majority of votes in five districts.

Both the 2004 and 2008 presidential elections pitted only two candidates in each election – one from the KMT and one from the DPP. In 2004, Chen sought re-election under the DPP endorsement as Lien Chan attempted to restore power to the KMT. In a very narrow and somewhat controversial election, Chen was able to capture a second term for the DPP, gaining 50.1 percent of the popular vote. He carried the same ten districts as he did in 2000 while also picking up Taichung County in central Taiwan. While obvious that Chen would win the majority of votes in those 11 districts, the margin of victory is noteworthy as he was either close to or exceeds a three-fifths majority in seven of those 11 districts. His largest margins of victory were in the south.

In 2008, Chen was unable to run again due to term limits. Therefore, the DPP nominated (Frank) Hsieh Chang-ting while the KMT nominated former Taipei mayor Ma Ying-jeou. The KMT was able to return to power with a landslide victory (58.45 percent versus 41.55 percent). Ma wrested away six districts and secured all seven municipalities. Hsieh's five victories were concentrated in the south.

Place, Identity, and Taiwan's Presidential Elections

The 12-year evolution from a seemingly 'solid Taiwan' to the cleavage in the south is rooted in both the historical experience of the island and the political dynamics of Taiwan.

It should first be noted that the appearance of a 'solid Taiwan' had more to do with Lee Teng-hui's overall popularity rather than an affinity towards the KMT or its traditional stance. Lee's approval rating remained above 60 percent. An ethnic breakdown of his approval ratings shows that he 'enjoyed rather high popularity among Taiwanese of Fujian origin, moderately high popularity among Hakka speakers, and rather low popularity among mainlanders' (Lu 2002, 62). It is important to note that his low approval rating among mainlanders, a stalwart of the KMT, demonstrates fractures in the base of the party. As an ethnic Taiwanese, he easily appealed to the 'native Taiwanese' that tend be pro-independence (Bedford and Hwang 2006). Furthermore, Lee was instrumental in deconstructing the old system and democratizing the island, something that was seen as a challenge to mainland China. Although it should be noted president Lee stated that he was not in favor of independence (Lu 2002).

The results of the 2000 election can be informed by the work of Webster (1992), who notes that the demise of the 'solid South' in the United States was due in part to the emergence of a third party. The emergence of a third party in Taiwan is crucial in allowing the opposition DPP to gain control of the Taiwanese presidency. James Soong's candidacy as an independent greatly divided both the mainlander vote, as well as the native Taiwanese vote. Cheng and Hsu (2002) note that the KMT nominee, Lien was not as popular as Soong. Lee, as chair of the

KMT, did not support the more popular Soong due to previous disagreements and chose the less popular Lien. As a result, Soong opted to run as an independent and immediately became the front-runner to win the election (Rigger 2001).

Because many KMT supporters did not feel that Lien would win the presidency, voters shifted their allegiance to the candidate best suited to defeat Chen – Soong. '[T]he KMT's Lien was the second preferred candidate for most voters, mainlanders and [native Taiwanese] alike. ... [C]onsistently lagging behind the other two contenders in polls rendered Lien a nonviable alternative, and eventually he became the victim of strategic voting' (Cheng and Hsu 2002, 166–7). The division of the KMT vote is important in that if we were to combine the votes for Lien and Soong, then the numbers would reflect previous elections in Taiwan (a 60/40 victory for the KMT). Thus, this division allowed for the DPP and Chen to narrowly wrest power away from the KMT.

Additionally, Webster (1992) suggests that the impetus for electoral change in the US South is situated in urban areas. While DPP candidate Peng Ming-min was unable to win a single district in the 1996 election, his highest shares were in metropolitan areas and provincial cities. Peng also did well in what would become staples of the DPP electorate – counties in the south as well as Yilan County in the northeast. Similarly in 2000 and 2004, in districts won by the KMT, Chen Shui-bian fared better in city districts than in county districts. Furthermore, DPP support tends to be the lowest in the rural counties of Hualien and Taitung. This pattern is reflected in numbers cited by Rigger (2001, 166), where she notes that DPP support is weakest among those involved in agriculture.

Another key point in the 2000 election is that Chen and the DPP distanced itself from the pro-independence platform. Chen quickly noted that the DPP only wanted a referendum on independence, rather than outright independence. For Chen's part, he moved his position more towards the middle, aligning himself more the with Lee's policies. While Lee Teng-hui never publicly endorsed Chen, many of those close to Lee backed the DPP candidate in the weeks leading up to the election. These endorsements included Lee Yuan-tseh, Noble laureate, and Hsu Wen-lung, president of plastic producer Chi Mei Corporation who claimed that Chen was best suited to 'continue Lee Teng-hui's course' (Rigger 2001, 193). These late endorsements gave Chen a final push to the presidency.

It is important to note that Chen's emphasis on self-determination was not simply political rhetoric during his campaign for presidency; Chen publicly espoused such views as early as 1991. While hardline leaders of the DPP wanted a platform seeking independence, Chen sought to soften the stance by stating that a move should be decided by the people and not by one political party. After a disastrous showing in the 1991 National Assembly elections, the DPP began moving further away from independence 'plank' and by 1996, once the hardliners left the DPP to form the Taiwan Independence Party, the DPP's position was closer to Lee Teng-hui's middle position. This stance placed the DPP in a more comfortable position to claim the presidency. Chen reiterated this stance in his 2000 inauguration speech in which he advocated the 'five noes' and the maintenance of the status quo (Chen 2000).[2]

While the DPP carries the mantle of the independence movement, in fact it has more in common with the KMT than public perception indicates. As Rigger (2001, 130) writes, 'the DPP still is no fan of unification or 'One China,' but it has accepted the two Chinas arrangement that exists, in effect, today.' This stance, which supports the status quo backed by the KMT, is promoted primarily because Taiwan is, for all intents and purposes, independent, but also because of security concerns. 'The DPP's current position is that because Taiwan already enjoys de facto independence, a declaration of *de jure* independence is unnecessary and would only antagonize Beijing.' Thus, DPP leaders 'accepted the logic that since Taiwan already enjoyed independence, there was no need to change the status quo' (128 and 130). Therefore, even though the DPP advocates for a referendum on Taiwanese independence (DPP website 2010), in general there is acceptance by the party of the status quo. It is through this 'new middle way' (Bedford and Hwang 2006) that Chen was able to calm fears of an independence push and work his way towards the middle.

The resulting victory by Chen and the DPP gave the opposition footing that carried over to the 2004 election in which Chen narrowly defeated a joint ticket of Lien and Soong. By 2008, however, the KMT was able to emphatically return to power as Ma Ying-jeou defeated DPP candidate Frank Hsieh. At this point, a clear geographical pattern emerged.

By taking the election results in terms of regional blocs (see Figure 6.1),[3] it is clear to see that the opposition DPP holds a firm grip on the south, while the KMT is strongest in the east and north (see Figures 6.2 through 6.5). The central districts in Taiwan do not clearly fall into one camp as it has voted for both KMT and DPP candidates. Also, there is one outlier as the DPP has claimed Yilan County in the north in two of the four elections – 2000 and 2004.[4]

So, what explains this pattern? There are several factors that come into play here. First, there is an ethnic difference. Rigger (2001, 164) notes that 'Taiwanese

2 While Chen did push for a referendum to rewrite the constitution, a move some argued was akin to declaring independence, he stated that a new constitution would not address independence. Chen, therefore, stood by his five noes and maintained the status quo.

3 While arbitrary, the author breaks down the regions of Taiwan as follows: Central (Changhua, Miaoli, Nantou, Taichung and Yunlin counties, and Taichung city), East (Hualian and Taitung counties), North (Hsinchu, Taipei, Taoyuan and Yilan counties, and Keelung, Hsinchu and Taipei cities), and South (Chiayi, Kaohsiung, Penghu, Pingtung and Tainan counties, and Chiayi, Kaohsiung, and Tainan cities). For this analysis, the island counties of Lienchiang and Kinmen were not included. These regions are consistent with other regionalizations of Taiwan.

4 Yilan County has demonstrated a long history of opposition to the KMT. It was one of the first to locally elect a non-KMT leader and, despite the 2008 election, remains a stronghold for the DPP (Wang 2009).

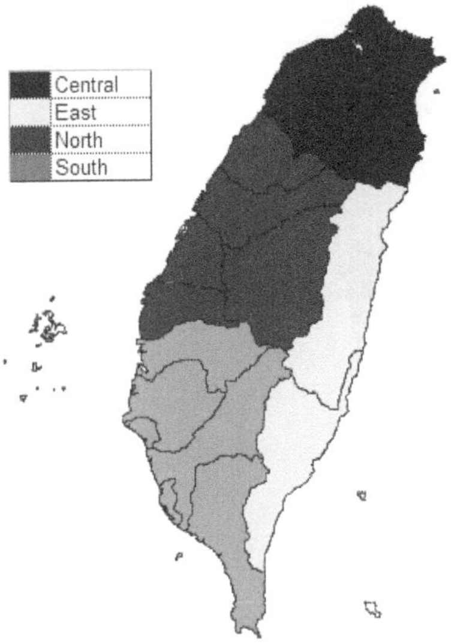

Figure 6.1 Map of the Regions of Taiwan

Source: Author.

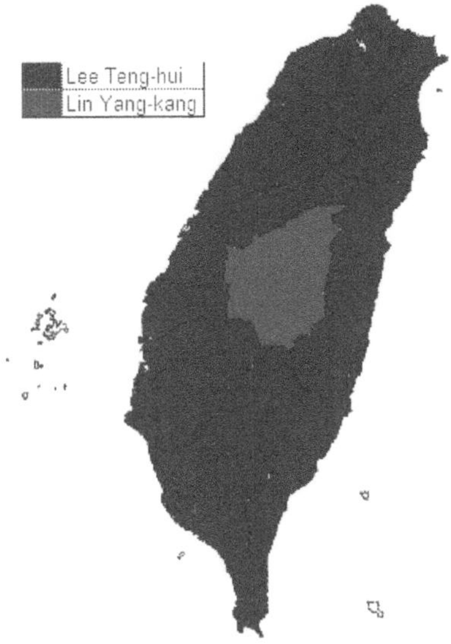

Figure 6.2 Map of the 1996 Presidential Election

Source: Author.

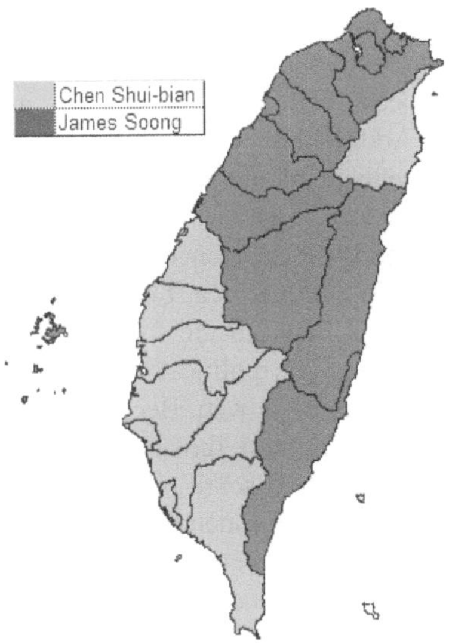

Figure 6.3 Map of the 2000 Presidential Election

Source: Author.

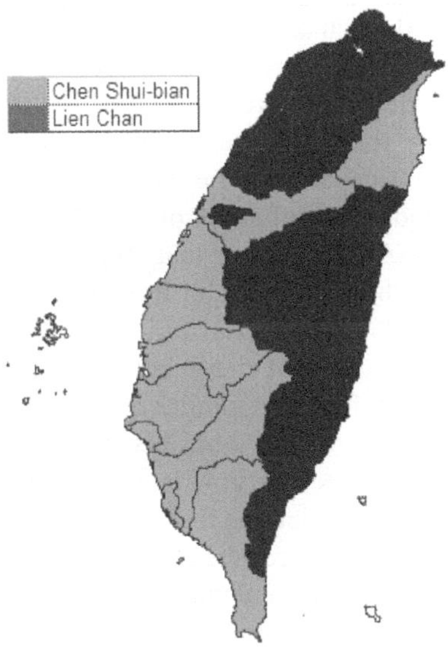

Figure 6.4 Map of the 2004 Presidential Election

Source: Author.

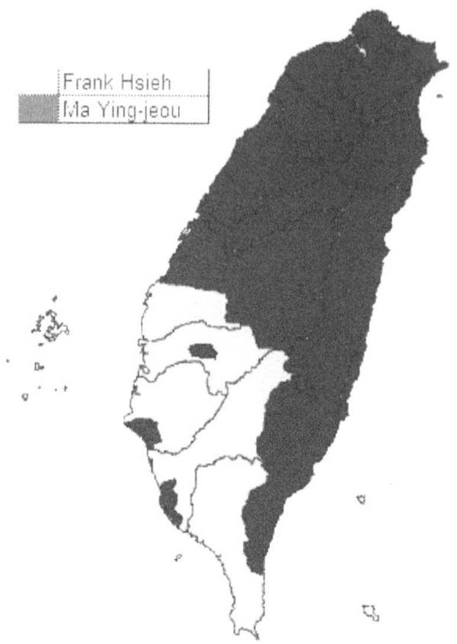

Figure 6.5 Map of the 2008 Presidential Election

Source: Author.

and Mainlanders were equally likely to support the KMT. However, Taiwanese were far more likely to support the DPP than were Mainlanders.' Historically, Mainlanders are concentrated in the north and east, both KMT strongholds, and are sparser in the south, the DPP's territory. Additionally, the eastern counties of Hualian and Taitung have the largest percentage of indigenous peoples, a traditional supporter of the KMT.

Another factor that Rigger (167) notes is 'political differences between the north and the south.' In their analysis of voting in South Korea, Lee and Brunn (1996) note that historical rivalries and resentment between the southwest and other parts of the country, especially the north, explain political cleavages in South Korea. A similar north–south rivalry is present in Taiwan. Because of this rivalry, 'the DPP has benefited from resentment in southern Taiwan at the wide discrepancy in government spending between north and south' (Rigger 2001, 167). Considering that the KMT long-controlled government spending, the south's affinity towards the opposition, in this case the DPP, makes sense.

But this resentment reflects the center of opposition that the south came to represent during the democratizing process and the emergence of (new) political parties. Rigger (167, emphasis added) notes that 'the DPP's emphasis on Taiwanese identity, including language and culture as well as Taiwan independence, has been well received in the south, *where Taiwanese consciousness is strong*.' Given that the KMT's platform with regards to mainland China is one of cooperation and reconciliation, the DPP is often held as the party of independence and therefore Taiwanese nationalism. This perception coupled with Taiwanese consciousness centers the south as a site of opposition.

The connection between the DPP and the south is galvanized by the Meilidao Incident, also known as the Kaohsiung Incident, which is held as one of the catalysts for the formation of the DPP. On December 10, 1979 in the southern city of Kaohsiung, a human rights protest was violently put down and three days later several political opposition leaders were arrested. The resulting outcry over this incident led to an increase in opposition movements, including a group of defense attorneys that included Chen Shui-bian. Seven years later, the Democratic Progressive Party was established in Taipei (Rigger 2001, Bedford and Hwang 2006). Although it was created in a hotel in the north, it was the incident in the south that fueled the opposition movement. The Meilidao Incident and its connection to the DPP places the opposition movement in the south.

What this all represents is that above any other indicator, place is a main factor in understanding the voting patterns in Taiwan. Agnew (1987, 34) writes that places can 'become collective actors.' This is important to note with regards to the role of political parties. 'The ideologies and policy proposals of different parties appeal more successfully in some places than in others as they strike responsive and nonresponsive chords among dominant sectors of the electorates.' Agnew continues by noting that certain actions, such as opposition movements, are rooted in places, which act as a seat of sentiment. In other words, a sense of place 'can present itself through different manifestations,' such as the DPP. 'Place, therefore, is not the *outcome* on a map of an abstract social process beyond place. Place refers to the *process* of social structuration' (36, emphasis in original). The pattern that takes shape in Taiwan is a result of place-based factors – concentration of mainlanders in the north; Kaohsiung as a symbol of opposition; KMT support among agricultural workers in the east – more so than simply a matter of ethnic difference or policy preference. Certainly the latter two variables are important, but each is rooted in space and helps create the place-based identity that gives rise to the pattern in Taiwan.

References

Agnew, J. 1987. *Place and Politics*. Boston: Allen and Unwin.

Bedford, O. and K.K. Hwang. 2006. *Taiwanese Identity and Democracy: The Social Psychology of Taiwan's 2004 Elections*. New York: Palgrave McMillian.

Booth, M. 1999. *The Dragon Syndicates: The Global Phenomenon of the Triads*. New York: Carroll & Graf Publishers.

Chen, S.B. 2000. Taiwan stands up: Toward the dawn of a rising era. [speech] Taiwan Public Policy Council. Transcript available at: http://www.tppc-us.org/LinPDF/ChenShuiBian5-20-00.pdf.

Cheng, T.J. and Y.M. Hsu. 2002. The March 2000 election in historical and comparative perspectives: Strategic voting, the Third Party, and the non-Duvergerian outcome. In B. Dickson and C.M. Chao (eds) *Assessing the Lee Teng-hui Legacy in Taiwan's Politics: Democratic Consolidation and External Relations*. Armonk, NY: M.E. Sharpe.

Cooper, J. 1998. *Taiwan's Mid-1990s Elections: Taking the Final Steps to Democracy*. Westport, CT: Praeger Publishers.

Cumings, B. 1984. The origins and development of the Northeast Asian political economy: Industrial sectors, product cycles, and political consequences. *International Organization*, 38 (1): 1–40

Democratic Progressive Party. 2010. Platform. Available at http://www.dpp.org.tw/index_en/.

Diamond, L. 2001. Anatomy of an electoral earthquake: How the KMT lost and the DPP won the 2000 presidential election. In M. Alagappa (ed) *Taiwan's Presidential Politics: Democratization and Cross-Strait Relations in the Twenty-first Century*. Armonk, NY: M.E. Sharpe.

Dickson, B. 1996. The Kuomintang before democratization: Organizational change and the role of elections. In H.M. Tien (ed.)*Taiwan's Electoral Politics and Democratic Transition: Riding the Third Wave*. Armonk, NY: M.E. Sharpe.

Hsieh, J. 2008. Democracy in a mildly divided society. In P. Paolino and J. Meernik (eds) *Democratization in Taiwan: Challenges in Transformation*. Burlington, VT: Ashgate Publishing.

Huang, T.F. 1996. Elections and the evolution of the Kuomintang. In H.M. Tien (ed.) *Taiwan's Electoral Politics and Democratic Transition: Riding the Third Wave*. Armonk, NY: M.E. Sharpe.

Huat, C.B. 2001. Defeat of the KMT: Implications for one-party quasi-democratic regimes in Southeast Asia. In M. Alagappa (ed.) *Taiwan's Presidential Politics: Democratization and Cross-Strait Relations in the Twenty-first Century*. Armonk, NY: M.E. Sharpe.

Kuomintang. 2010a. 'Introduction.' [Online]. Available at http://www.kmt.org.tw/english/page.aspx?type=para&mnum=105.

Kuomintang. 2010b. 'History.' [Online]. Available at http://www.kmt.org.tw/english/page.aspx?type=para&mnum=108.

Lee, D.O. and S. Brunn. 1996. Politics and regions in Korea: An analysis of the recent presidential election. *Political Geography*, 15(1): 99–119.

Lu, Y.L. 2002. Lee Teng-hui's role in Taiwan's democratization: A preliminary Assessment. In B. Dickson and C.M. Chao (eds) *Assessing the Lee Teng-hui Legacy in Taiwan's Politics: Democratic Consolidation and External Relations*. Armonk, NY: M.E. Sharpe.

Myers, R., L. Chao and T.C. Kuo. 2002. Consolidating democracy in the Republic of China on Taiwan, 1996–2000. In B. Dickson and C.M. Chao (eds) *Assessing the Lee Teng-hui Legacy in Taiwan's Politics: Democratic Consolidation and External Relations*. Armonk, NY: M.E. Sharpe.

Rigger, S. 2001. *From Opposition to Power: Taiwan's Democratic Progressive Party*. Boulder, CO: Lynne Rienner Publishers.

Tien, H.M. 1996. Elections and Taiwan's democratic development. In H.M. Tien (ed.) *Taiwan's Electoral Politics and Democratic Transition: Riding the Third Wave*. Armonk, NY: M.E. Sharpe.

Webster, G. 1992. Demise of the solid South. *Geographical Review,* 82(1): 43–55.

Wright, M. 1955. From revolution to restoration: The transformation of Kuomintang ideology. *The Far Eastern Quarterly*, 14(4): 515–32.

Wu, J. 1995. *Taiwan's Democratization: Forces behind the New Momentum*. Hong Kong: Oxford University Press.

Chapter 7
The Power of Definition: The Puerto Rican Diaspora and the Island's Status Referenda

Luis D. Sánchez-Ayala

Almost every Puerto Rican agrees that sports raise passions and wake profound sentiments among their compatriots. The list of those sports so highly regarded includes basketball, baseball, boxing, and of course, at the top of the list, voting in any electoral event. For Puerto Ricans, the electoral debate is present in everyday popular discourse and in the spaces of everyday life. Politics is the main subject discussed at the coffee shop, in the beauty parlor, or while waiting in line at the supermarket or the bank. Also, often Puerto Ricans choose the colors of their houses, cars, or even their clothes according to their electoral preferences.

This is the reality of a country whose political future is still uncertain after centuries of foreign domination through a variety of colonial arrangements. Consequently, for Puerto Ricans any electoral event in the island has a potential to influence the political future and final status of the nation. However, Puerto Ricans have been unable to agree on one particular status formula to solve the island's political dilemma; they continue to disagree over which direction would be best to follow.

There have been three popular referenda on the island offering a variety of formulas as options for a final political status of Puerto Rico. Those referenda were celebrated in 1967, 1993, and 1998, but none has succeeded in solving the political limbo of the island.

Since the last referendum in 1998, many more initiatives to celebrate another referendum in the island have been discussed in both Washington and San Juan, but the debate regarding Puerto Rico's political status referenda is much more complex than this would suggest. This complexity exists because the universe of Puerto Ricans interested and involved in this matter expands well beyond the territorial boundaries of Puerto Rico. Currently there are nearly five million self-identified Puerto Ricans living in the United States, many of whom are still attached to the political issues of the island. However, these members of the Puerto Rican diaspora are not officially authorized to vote on any electoral event in the island. Yet discussions about whether Puerto Ricans living in the United States should be included in any consideration regarding the final decision about the island are taking place on the island and in the United States. The controversy rests on who would be considered "Puerto Rican" for the purpose of participating in any potential status referenda.

This twist gives a new significance to the debate over Puerto Rico's status situation. Not only is the political situation of the island being deliberated, but there also debates about what it means to be Puerto Rican among members of the Puerto Rican diaspora. In this regard, determining who can vote or participate in any decision-making process dealing with the island's political future implies issues of identity definition and construction for the nearly five million Puerto Ricans who live in the United States. Nonetheless, in the case that Puerto Ricans living in the United States are included in any potential status referendum, they will represent a significant force in deciding the political future of Puerto Rico and the Puerto Ricans living in the island.

The Status Debate

The debates over the political status of Puerto Rico unfold on two different fronts, those that originated in Puerto Rico and those that originated in Congress or within the United States government.

The status debate in Puerto Rico is present in the everyday popular discourse of the island. The entire political system is arranged around the status issue. The three existing political parties in the island were founded to respond to the status issue, each one representing a different status option: statehood, commonwealth, and independence.

The debates originating in Puerto Rico have led to referenda on the issue on three different occasions. Those referenda were held in 1967, 1993, and 1998 by initiative of the government of Puerto Rico in agreement with the three local political parties (the pro-statehood New Progressive Party, the pro-commonwealth Popular Democratic Party, and the pro-independence Puerto Rican Independence Party). However, in all three cases the referenda have not succeeded in their goal of solving the political limbo of the island and have not been approved or endorsed by the government of the United States. Without Congressional approval or endorsement, the referendum results were reduced to popularity contests between the three political parties in the island.

Nonetheless, lack of attention from Congress was not the only obstacle to the effectiveness of these referenda. Out of the three referenda, two were boycotted by at least one of the political parties on the island. In 1967 the Puerto Rican Independence Party, known by its Spanish-language acronym, PIP, boycotted the referendum, arguing that its only purpose was to reinforce the Commonwealth image in front of the international community. On the other hand the Statehood Republican Party, which at that time was the pro-statehood party, was divided, and although its leaders supported the boycott, a group of the party's young leadership led a movement to participate in the referendum.

The other referendum that was boycotted was the one held in 1998. This referendum was initiated by the governing New Progressive Party, known as

"PNP," the successor of the Statehood Republican Party), and supported by the Puerto Rican Independence Party. The referendum included four options:

1. Commonwealth status, or the application of the US Congress sovereignty over Puerto Rico;
2. Free Association, a non-colonial association with the US;
3. statehood, the integration of Puerto Rico into the union as the 51st state; and
4. independence, the recognition of Puerto Rico as a free and sovereign country.

The pro-commonwealth Popular Democratic Party (PPD), which never has initiated any debate or conversation about the political status issue, opposed the referendum. The leaders of the party did not agree with the PNP and the PIP on the definition that was given to the term "Commonwealth." As a result, the three parties agreed to include a fifth column that would say "None of the Above." The PPD then started a campaign promoting "None of the Above," arguing that the referendum was not legitimate.

In the 1967 referendum, the commonwealth option prevailed with 60 percent of the votes, against 39 percent for statehood and one percent for independence (Figure 7.1). The Commonwealth option dominated almost every municipality of the Island, obtaining more than 50 percent of support in most. However, the Puerto Rican Independence Party's boycott and the Statehood Republican Party's internal division set the stage for the victory of the commonwealth option and its promoter, the PPD.

In the early 1960s the island was still dominated by a strong PPD. Just a decade after the creation of the Commonwealth of Puerto Rico, the party (main actor in the creation of the commonwealth status) was in the last stage of its electoral momentum. Its involvement in the referendum had a deep effect on the population's voting behavior.

The 1993 referendum showed a changing pattern. Even though the commonwealth option prevailed, its margin of support had decreased dramatically since the 1967 referendum. On the other hand, support for statehood increased significantly, obtaining significant support. Likewise, the difference in support percentages between commonwealth and statehood grew closer in comparison to the 1967 referendum.

In the case of independence, there were significant changes. Support for independence increased significantly, from barely one percent to four percent, although it remained a small minority. In Puerto Rico, both the local government and US Government have often condemned those who support the independence option. Systematic vigilance against and repression of independence supporters were common practices, applied simultaneously with an exaggerated and unrealistic fear campaign conducted among the population against independence. The fact that the independence movement has resisted and survived all those obstacles for so long is a demonstration of its strength.

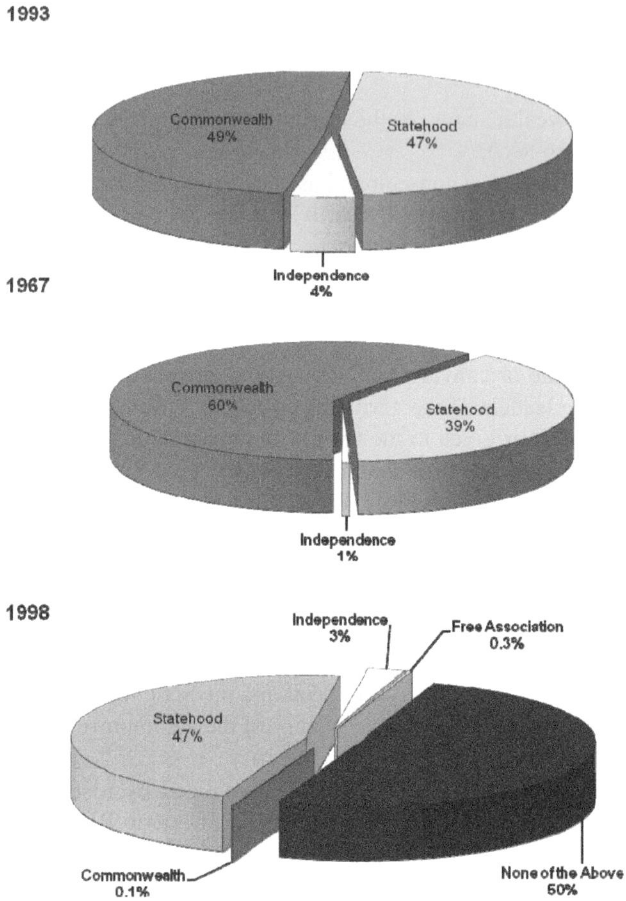

Figure 7.1 Status Referendums in Puerto Rico
Source: Comisión Estatal de Elecciones de Puerto Rico.

Finally, the 1998 referendum deviated from the electoral patterns found in the previous two referenda. Held several months after hurricane George devastated the island, and with the strong opposition of the PPD, the population was not enthusiastic about the election. The low level of enthusiasm was demonstrated by the referendum's low turnout.

In this referendum, 50 percent of the voters voted for "None of the Above," a fifth column championed by those who were opposed to the celebration of the referendum. The PPD made a call for voters to select "None of the Above," calling it "the option of the people." In the commonwealth-statehood-independence referenda, the PPD, who opposed to the Commonwealth definition on the ballot, managed to divert the debate and change the referendum into a punishment vote for the government. Regardless of this, the statehood option maintained its support (47 percent). Likewise, independence got three percent, commonwealth (as appeared in the ballot), 0.1 percent, and the fifth column "None of the Above" obtained 50 percent, just one percent more than the commonwealth option obtained in the 1993 referendum.

Most choices in politics are framed in terms of a new proposal being pitted against the status quo, and in such a situation the status quo is at an advantage (Christin, Hug, and Sciarini, 2002: 764). In Puerto Rico, the commonwealth option has an advantage for those who are uncertain of whether the new political status options will achieve their goals. However, there is enough evidence to state that political status preferences are currently changing.

The results of the three referenda demonstrated the shift of the voting behavior in Puerto Rico. Currently the popularity of the commonwealth option is in decline, while support for statehood has been rising. Support for remaining a commonwealth has gone from 60 percent in 1967 to 0.1 percent in 1998, while support for statehood has gone from 39 percent in 1967 to 47 percent in 1998. Likewise, support for independence went from one percent in 1967 to 3 percent in 1998. But if we take out the 1998 results, since the commonwealth supporters boycotted the commonwealth option and instead voted for the "None of the Above" option, the same pattern can be seen. Also, if we add the percentage of support for None of the Above (50 percent) with the percentage in favor of the commonwealth (.1 percent), it would not represent a significant difference. Overall, support for the commonwealth is still showing a decline, contrary to the increase in support for statehood.

These patterns can also be seen in the results of the general elections. In Puerto Rico, the elections are intimately related to the political status discussion. The role of the three political parties is to promote a political status formula (statehood, commonwealth, or independence). The general elections are dominated by status issues, and the status referendum campaigns in turn are dominated by the political parties. As LeDuc (2002) stated, "the positions that political parties take, either in the evolution of the referendum issue or during the campaign itself, provide one of the strongest available information cues to voters."

In 1989, the heads of the three political parties in Puerto Rico (among them the governor of Puerto Rico) wrote a letter to President Bush demanding a response to the island's status situation. As a result, President Bush instructed Congress to take action. Senator Bennett Johnston (D-Louisiana) presented in the Senate the S. 712 "Puerto Rico Status Referendum Act." The bill was referred to a Senate Committee, where it died in 1991; it never made it to the floor to be considered and voted on.

In 1996, the Republican leaders of the House, through the chair of the Resources Committee, Don Young (R-Alaska), presented H.R. 856, the "United States-Puerto Rico Political Status Act." The House approved the bill by one vote, after an overnight session and a heated debate. However, the bill was not passed by the Senate; it did not even make it to the floor for consideration, as was the case for the S. 712 in 1989.

More recently, discussions over the political status of Puerto Rico have taken place in Congress. This time Congressional action has responded to the White House's report released in December 2005 by the President's Task Force on Puerto Rico's Status. Following the report's recommendations, Congress is debating possible legislation to address the political situation of Puerto Rico.

Soon after the White House's report was released, the political parties in Puerto Rico started to pressure Congress. Consequently, two bills were introduced in Congress, one in the Senate and one in the House. The bill introduced in the Senate, designed and supported by the pro-commonwealth party in Puerto Rico, discards the White House's recommendations of a referendum only offering statehood and independence. This bill does not contemplate a referendum or any further debate over Puerto Rico's stature. Rather, it proposes that Puerto Ricans should first decide internally which path they want to follow. On the other hand, the bill introduced in the House has been designed by the pro-statehood party with the support of the pro-independence party. This bill follows the recommendations of the Presidential Task Force on Puerto Rico's Status, so it proposes holding a referendum on the island offering statehood or independence. The Senate bill died soon after its introduction; however, the House bill H.R.2499, the "Puerto Rico Democracy Act of 2010" passed on April 29, 2010 and now must be taken up by the Senate. This bill would introduce a two-step ballot measure for Puerto Rico to decide if Puerto Ricans want to change their current relationship with the United States. If they vote to change their status, they can then choose to become a state, pursue independence, or seek some other "political association between sovereign nations."

The Puerto Rican Diaspora

Over the past century, millions of Puerto Ricans have migrated to the United States. The 2008 American Community Survey conducted by the US Census Bureau estimated the population of Puerto Rico at 3,954,037 and the number of Puerto

Table 7.1 Puerto Rican Population in the Island and the United States

Year	Puerto Rico	United States
1940	1,869,255	69,967
1950	2,210,703	301,375
1960	2,349,544	887,662
1970	2,712,033	1,492,396
1980	3,196,520	2,013,945
1990	3,522,037	2,727,754
2000	3,808,610	3,406,178
2008	3,954,037	4,216,533

Source: US Census Bureau.

Ricans living in the US at 4,216,533 (Table 7.1). According to these numbers, currently there are more Puerto Ricans living outside of Puerto Rico than there are living in the island. This situation has given rise to a diasporic community larger than the total population of the island. Therefore, defining Puerto Ricanness is a process that reaches far beyond the island's boundaries.

Puerto Ricans have been migrating to the mainland since the island became a possession of the United States in 1898. By the first decades of the 1900s, Puerto Ricans were being hired to work as seasonal workers as far away as Hawaii (Rosario 1983; Maldonado-Denis 1980; Mintz 1955). The 1910 US census estimated the Puerto Rican population on the mainland at 1,500. This figure increased even more after the approval of the Jones Shafroth Act of 1917, which established the island as an organized but unincorporated territory of the United States and granted US citizenship to Puerto Ricans. This act facilitated the movement of labor to the mainland, significantly raising the net migration from the island to the mainland in the 1920s. However, the peak period of migration began with the end of World War II. During this period, which became known as the Great Migration, the Puerto Rican community grew to more than half a million.

Nonetheless, the Great Migration led to the creation of Puerto Rican barrios in New York City's South Bronx, Spanish Harlem, Manhattan's Lower East Side, and Brooklyn's Atlantic Street. These barrios began to resemble "Little Puerto Rico," characterized by Puerto Rican ways and customs.

During the 1960s and 1970s, the migratory flow from the island to the mainland took other forms. At this time a drastic decrease in the outflow of migrants from the island to the mainland and a rapid increase of return migration to Puerto Rico occurred. Consequently, beginning in 1965, Puerto Rican migration was characterized by continuous back and forth movement between the island and the mainland (Lorenzo-Hernández 1999: 989).

From the 1980s to the present, an increase in migration to the United States has been taking place. The new migrants have demonstrated dispersion patterns outside of the northeast of the United States, reaching to the South and West in states like Florida, California, and Texas (Figure 7.2).

In the 1980s, over 400,000 Puerto Ricans left the island for the mainland. In fact, 12.3 percent of the Puerto Rican population that resided on the island in 1980 was no longer living there in 1990 (Rivera-Batiz and Santiago 1990). Puerto Ricans' net migration to the mainland increased during this time period, reaching 126,465 in the 1980s and 111,336 in the 1990s. The combined total of these numbers represented over a quarter of a million people, or more than 7 percent of the island's population in 1980 (Marquez 2004).

From 1990 to 2000 the Puerto Rican population in the United States increased faster than the population on the Island. Between 1990 and 2000, the population of mainland Puerto Ricans grew by 25 percent while the population in the island increased by approximately 10 percent (Marquez 2004). The 2008 US Census Bureau American Community Survey estimated the island population at 3,954,037

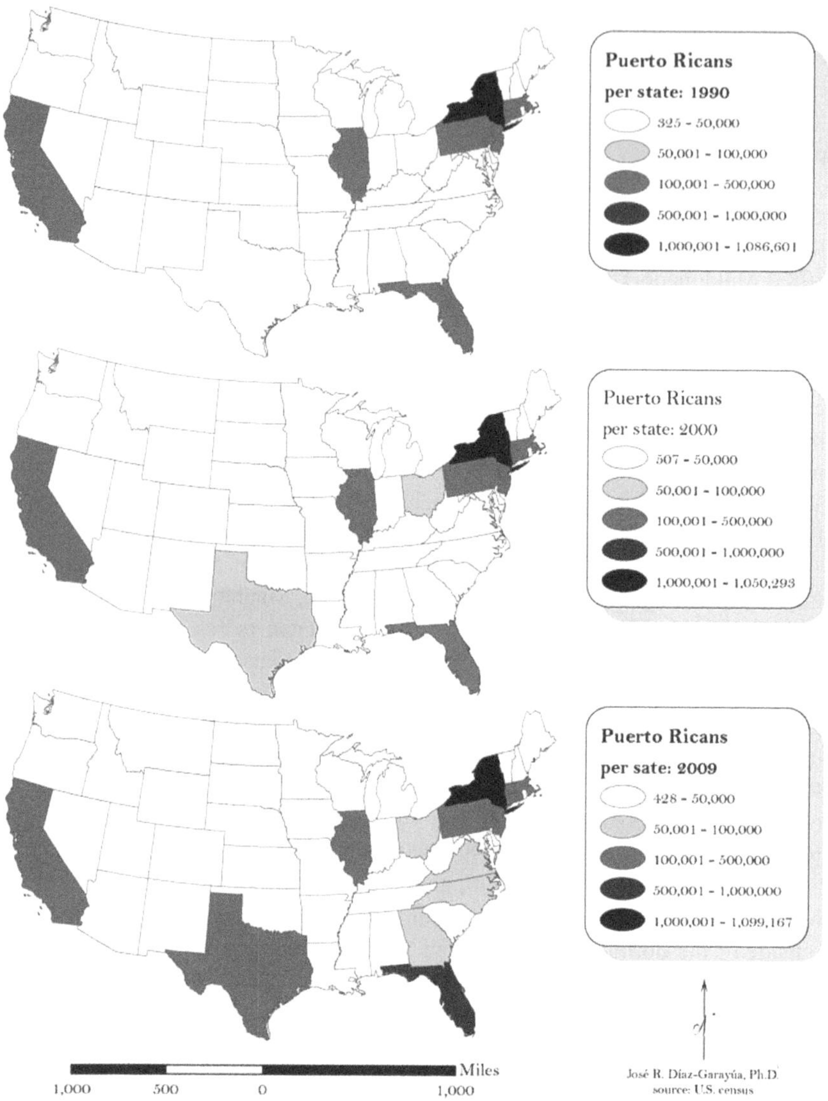

Figure 7.2 Distribution of Puerto Ricans in the United States, 1990–2009
Source: Author.

and the number of mainland Puerto Ricans at 4,216,533: for the first time there are more Puerto Ricans living in the mainland United States than on the island.

Historically, Puerto Ricans in the United States concentrated in New York, New Jersey, Pennsylvania, Ohio, Illinois, and other Northeastern and Midwestern states. However, during this last emigration wave the mainland Puerto Rican population started to disperse much more widely than in the previous emigration movements. Consequently, the proportion of Puerto Ricans significantly increased in other states such as Florida, which has had the greatest increase, and Texas, and California. Yet regardless of the fact that New York still has the largest percentage of Puerto Rican residents in the US, the growth of Florida's Puerto Rican population has been spectacular, from slightly more than two percent of all stateside Puerto Ricans in 1960 to more than 17 percent by the year 2006. These numbers placed Florida with the second largest concentration of Puerto Ricans in the United States and as the fastest growing Puerto Rican community.

Status Referenda and Puerto Ricanness

The political status debate over Puerto Rico's political situation, besides dealing with the issue of defining the island's political future, had been broadened to include issues of Puerto Rican identity in Puerto Rico and the diaspora. This is a particularly important issue as identities are more than just labels; identities also involve people's acceptance, resistance, choice, particularity, invention, perceptions, rejections, and so forth (Cornell and Hartmann 1998: 77). Therefore, identities are socially constructed through structural processes rather than being innate or pre-given (Bondi 1993: 86). This process gives meaning to identities on the basis of cultural attributes. Identity is a product, and, at the same time, the producer of social phenomena; it is always an incomplete process. Accordingly, Keith and Pile (1993: 28) suggest that "identity formation is endless because it is based on a moment of arbitrary closure which in the same fashion is both true and false simultaneously." In that sense the concept of identity is a highly contested one.

Puerto Ricans, regardless of their location inside or outside of Puerto Rico, show a strong sense of national identity. However, when it comes to defining their community, each Puerto Rican has his or her own interpretation and meanings (Sánchez 2009: 66). In that sense, when it comes to identity, each individual potentially experiences a unique understanding of his or her surroundings, and this understanding is shaped by mental processes of information gathering and organization that make self-consciousness (Gold 1980). Geographers such as Peet (1998) argued that the spatiality of everyday life generates fundamental self-consciousness.

However, the significance of all this is the fact that the debate over Puerto Rico's political situation is put in motion a process of identity construction for Puerto Ricans in Puerto Rico, but equally importantly, in the United States. In this context, the discussions taking place in the US Congress about whether Puerto Ricans living in the United States should be included in any consideration regarding the final decision about the island are increasingly dominating the status debate. The controversy rests on how and who would be considered Puerto Rican in order to participate in any potential status referendum. Therefore, this controversy could mean profound changes in the way many Puerto Ricans living in the United States have constructed their identities.

That congressional debate over the political future of Puerto Rico has many angles that directly affect the Puerto Rican diaspora. Perhaps the most prominent ones can be summarized in the views of Congressional Representatives José Serrano from New York and Luis Gutiérrez from Chicago. Serrano (born in Puerto Rico) has argued that all persons who were born on the island should participate in any potential mechanism to solve Puerto Rico's status issue. On the other hand, Gutiérrez (born in the United States, first generation of Puerto Rican immigrants) maintained that all persons of Puerto Rican descent should be included in any decision regarding the political future of Puerto Rico. Along with these views there are many others that range from allowing participation to anyone claiming to be Puerto Rican to only including the residents of the island in such a decision.

Each possible scenario has different implications for the identity-making process of the Puerto Rican diaspora. If Congress chooses to follow the interpretation of Serrano, affirming that all persons who were born on the island should participate, Congress will then determine that a Puerto Rican is only a person born in the island. Therefore, in that sense the message to the diaspora would be that Puerto Ricanness is restricted to place of birth and any other individual, regardless of his or her ancestry, would not be recognized as such. This scenario would drastically affect the way that millions of "Puerto Ricans" born in the United States produce and reproduce their individual and collective identities.

On the other hand, if Congress chooses to favor the views of Luis Gutiérrez, affirming instead that the participation in a status referendum should be open to all persons of Puerto Rican descent, the scenario for the Puerto Rican diaspora would be different. In this case, anyone who can trace his or her ancestry to the island would be eligible to vote, and thus included within the boundaries of Puerto Ricanness. However, this interpretation could open the door for the question of where to put a limit, or if there should be a limit for such claim of ancestry. This flexible or limitless perspective could as well affect the way that the Puerto Ricans in the diaspora produce and reproduce their Puerto Ricanness, as it could stretch to the limit the particularities that distinguish a migrant group from a diaspora.

Similarly, a congressional decision to restrict the participation on a potential referendum to only residents of the island could influence some perceptions of Puerto Ricanness among the Puerto Rican diaspora. In this sense, such exclusion could draw a boundary of division/distinction that could be interpreted as the legitimation of the true Puerto Ricans, in opposition to the "others." Consequently, this would mean a potential process of identity [re]definition for the Puerto Rican diaspora.

These different points of view on the status debate over Puerto Rico demonstrate the complexity behind the consideration of a referendum in Puerto Rico. Yet, regardless of which direction the debate will finally take, Congress will be deciding much more than the political future of Puerto Rico; they will as well be setting the guidelines of Puerto Ricanness. Thus, in the case of Puerto

Rico, Puerto Ricans would not be only voting to decide the political future of the island, they will also be defining the boundaries of their community, in both the island and the diaspora. Therefore, a status referendum in Puerto Rico could put in motion a process of identity construction for Puerto Ricans in Puerto Rico, but more importantly in the United States.

For Puerto Ricans living in Puerto Rico, considerations of their political status could lead them to a process of defining their Puerto Ricanness in relation to the political status of Puerto Rico (e.g., Puerto Rican, Puerto Rican-American, American, etc.). However, for the Puerto Rican diaspora the answer to the question of whether they should be included in any consideration regarding the final decision about the island is particularly important as their identity as Puerto Ricans could be dramatically [re]defined with the outcome of the congressional debate over a potential status referendum for Puerto Rico. Therefore, for the Puerto Rican diaspora, voting in such a referendum would mean much more than just participating in an electoral event; it could instead signify the way in which they make sense of themselves individually and collectively, and thus, it could determine the future of their community.

Moreover, the status debate is making visible notions of difference within the Puerto Rican nation. While Puerto Rican identities are based on language, race, ethnicity, religion, sexuality, gender, class, nationalism, they are predominantly based on place of origin (Sánchez 2009: 7). Therefore, answering the question of who is Puerto Rican or what makes someone Puerto Rican, to then determine who would be included for the purpose of participating in any potential status referendum, has had the effect that terms such as "island-born Puerto Rican," "mainland-born Puerto Rican," "Nuyorican," "Orlandorican," and "Diasporican" are now being used as the main means to define affiliation. For that purpose, Puerto Ricans are creating (or reviving) labels and concepts of "othering." The labels used to differentiate social groups are important for the psychological segregation and perpetuation of the categories (Lorenzo-Hernández 1999: 991). In this sense, the consequential [re]definition of Puerto Ricanness is establishing opposites such as "us versus them." We must not forget that one of the most powerful ways to construct an identity is by identifying against (Cornell and Hartmann 1998). First define the "other" and then we define ourselves as "not the other," the opposite or what the other is not. But in the same way we define "others," they may also define us, and their definition may not match our own self-identity (Sánchez 2009: 8).

The status referendum debate is making Puerto Ricans aware of the fact that Puerto Ricanness means different things for different people at different places in different times. Consequently, such differences are being used to build boundaries between different notions of Puerto Ricanness, and thus to determine who can vote or participate in any decision-making process dealing with the island's political future. As Barth (1996: 15) puts it:

> the dichotomization of others as strangers, as members of another ethnic group, implies recognition of limitations on shared understandings, differences in

> criteria for judgment of value and performance, and a restriction of interaction to sectors of assumed common understanding and mutual interest.

The status debate is also accentuating the notions of difference among members of the Puerto Rican diaspora. In response to all the discussion and questions that the status debate had initiated, the members of the Puerto Rican diaspora are trying to identify among them who are the "real" Puerto Ricans. Consequently, the Puerto Rican diaspora, and their notions of difference, is splitting between the "old" communities (particularly the three historical communities of New York, Chicago, and Philadelphia) and the "new" communities (particularly central Florida).

Puerto Ricans in central Florida are distancing themselves from the rest of the diaspora, also adopting the use of terms such as "Nuyoricans" to reinforce their authenticity claims. They are building an identity based on what they think is from the island, away from the rest of the Puerto Rican diaspora on the mainland (Sánchez 2009: 92). On the other hand, Puerto Ricans in the historical communities perceive themselves as the center of the Puerto Rican diaspora, and thus, an integral part of the Puerto Rican nation, rejecting any claim of distinction from those in other parts of the United States.

Equally important in this debate is the fact that a possible vote of the Puerto Rican diaspora in a status referendum over the political future of the island could represent the difference between which option ultimately wins. The current volatility in Puerto Rico regarding popular support of the different political status options shows that no status option has sufficient support in the island to be considered representative of the political desires of the people of Puerto Rico as a whole. The results of all three referenda on the island have shown how supporters of commonwealth and statehood are merging together, with almost no difference between the percentage of Puerto Ricans supporting the commonwealth option and those supporting statehood supporters for the two options. Therefore, since Puerto Rican residents on the island are nearly evenly divided on their political desires, the votes of the Puerto Rican diaspora could decide which political status option would win and become the permanent political status of Puerto Rico.

The Puerto Rican diaspora is a significant element in the discussions over a potential vote to finally solve the political limbo of Puerto Rico. Yet, they are significant in the debate not only because their potential vote could affect the ultimate outcome of such a referendum, but more importantly because the consequences of the decision of whether or not to include them in such a referendum to begin with will cause reactions that could [re]define Puerto Ricanness, both in Puerto Rico and the United States.

Conclusion

The debate over the possible inclusion of members of the Puerto Rican diaspora in a referendum regarding the political future of the island gives a new significance to the debate over Puerto Rico's status situation. The debate has the potential to put in motion a process of identity definition among the Puerto Rican nation. Consequently, not only the political situation of the island is being deliberated, but also the question of who is Puerto Rican or what makes someone Puerto Rican, particularly among the Puerto Rican diaspora.

Determining who can vote or participate in any decision-making process dealing with the island's political status situation implies issues of identity definition and construction. The outcome of those considerations will potentially [re]define Puerto Ricanness, particularly for the nearly five million Puerto Ricans who live in the mainland United States. In this sense, the making of Puerto Ricanness blends with the definition of the political future of the island.

In the case of the diaspora, the process to find an answer to the question of who is Puerto Rican or what makes someone Puerto Rican, to then determine who would be included for the purpose of participating in any potential status referendum, is initiating a process of self-definition, in which Puerto Ricans are constructing their identities in an inbetweenness of being "here" but also being "there." This stage of in-betweenness, or "third space" (Bhabha 1994; Soja 1996), is a state of belonging to both and to neither, or as Bhabha (1994: 219) said, "neither one not the other but something else besides, in-between." This diversity allows the multiplicity of identities based on the existence of multiple layers of identity (Sánchez 2009: 72).

Determining who can vote or participate in any decision-making process dealing with Puerto Rico's political future not only had initiated a process of identity definition and construction for Puerto Ricans in Puerto Rican and the United States, but has also demonstrated that the task of identifying Puerto Ricanness is particularly difficult.

References

Bhabha, H. 1994. *The Location of Culture*. London: Routledge.

Bondi, L. 1993. Locating identity politics. In M. Keith and S. Pile (eds) *Place and the Politics of Identity*. London: Routledge, 84–101.

Brah, A. 1996. *Cartographies of Diaspora: Contesting Identities*. London: Routledge.

Christin, T., S. Hug, and P. Sciarini. 2002. Interests and information in referendum voting: An analysis of Swiss voters. *European Journal of Political Research* 41: 759–76.

Cornell, S. and D. Hartmann. 1998. *Ethnicity and Race: Making Identities in a Changing World.* Thousand Oaks, CA: Pine Forge Prints.

Gold, J. 1980. *An Introduction to Behavioural Geography*. Oxford: Oxford University Press.

Keith, M. and S. Pile. 1993. The place of politics. In M. Keith, and S. Pile (eds) *Place and the Politics of Identity*. London: Routledge, 22–40.

LeDuc, L. 2002. Opinion change and voting behaviour in referendums. *European Journal of Political Research* 41: 711–32.

Lorenzo-Hernámdez, J. 1999. The Nuyorican's dilemma: Categorization of returning migrants in Puerto Rico. *International Migration Review* 33(4): 988–1013.

Maldonado-Denis, M. 1980. *The Emigration Dialectic: Puerto Rico and the USA.* New York: International Publishers.

Márquez, C. 2004. Adios, Puerto Rico; Hola, United States. *Caribbean Business* September 16.

Mintz, S. 1955. Puerto Rican migration: A three-fold comparison. *Social and Economic Studies* 4: 311–25.

Peet, R. 1998. *Modern Geographical Thought*. Oxford: Blackwell

Rivera-Batiz, F. and C. Santiago. 1996. *Island Paradox: Puerto Rico in the 1990s*. New York: Russell Sage Foundation.

Rosario, C. 1983. *Éxodo Puertorriqueño: Las Emigraciones al Caribe y Hawaii, 1900–1915*. San Juan: Ramallo Brothers.

Sánchez, L. 2009. *The New Puerto Rico?: Identity, Hybridity and Transnationalism within the Puerto Rican Diaspora in Orlando, Florida*. Germany: VDM Verlag.

Soja, E. 1996. *Third Space: Journeys to Los Angeles and Other Real-and-Imagined Places*. Cambridge, MA: Blackwell.

PART III
Electoral Geography in the US

Chapter 8

Class, Ethnicity, Religion and Place in the 2008 US Presidential Election

Barney Warf

The 2008 presidential election was unquestionably "as riveting and historic a spectacle as modern politics has ever produced" (Heilemann and Halperin 2010, ix). After eight years of Republican rule under George W. Bush, a period that witnessed two foreign wars and rising economic dysfunction, the United States elected its first African-American leader, Barack Obama. By a margin of 53 to 46 percent of the votes cast (69.3 million v. 59.6 million), and 365 to 174 in the electoral college, Obama won an election that is by any measure historic.

As always, this contest played out unevenly over space, as local constellations of economic, political, and cultural forces favoring Obama and his GOP opponent, John McCain, were precipitated in different combinations and to varying

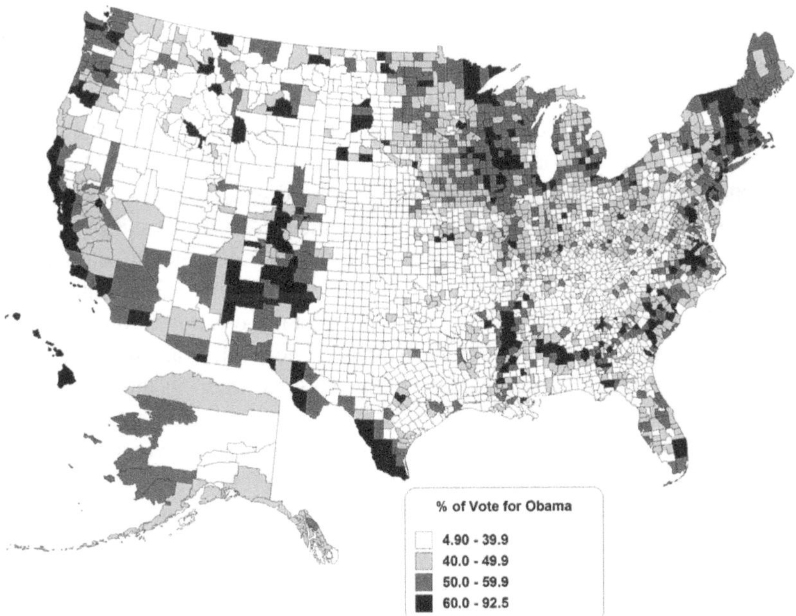

Figure 8.1 Percent of Votes for Obama by County, 2008

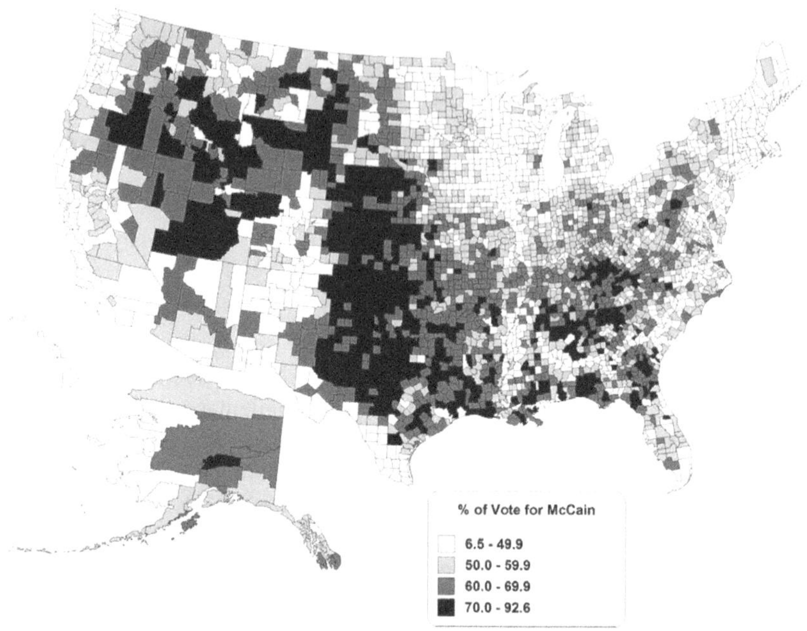

Figure 8.2 Percent of Votes for McCain by County, 2008

levels across the face of the nation. Obama's primary sources of strength lay in metropolitan areas, the West Coast, the upper Midwest, New England, and a belt of historically and predominantly African-American counties in the South (Figure 8.1). For the first time in decades, traditionally Republican-leaning states such as Virginia, North Carolina and Indiana joined the Democratic column. Conversely, McCain garnered majorities in most rural areas and small towns, notably in much of the South, the Great Plains, and the intermontane West (Figure 8.2).

However, underlying this geography are several sociospatial categories that profoundly shape American electoral landscapes, notably class, ethnicity, and religion. This chapter examines the 2008 presidential election through the lenses of each of these dimensions. The first three sections offer an overview of how each of these categories has shaped voting patterns, often in unexpected ways. Class, for example, traditionally absent in American politics, is not manifested in a simple dichotomy of wealthy support for Republicans and lower income support for Democrats, but rather the reverse; ethnic support for different political parties is widely variable among groups, and often within them; and the influence of religion varies markedly by denomination. The fourth section offers a quantitative analysis using geographically weighted regression, a means of examining the spatially uneven strength of different variables' influence on votes for Obama while statistically controlling for the influence of confounding forces. The results are presented in a series of maps. The concluding discussion links these outcomes to a wider discussion of the geographies of identity, difference, and marginalization.

Class, Income, and Party Alignment: A Counterintuitive Pattern

Simplistic views of American politics tend to assume that the Republican Party represents vested corporate and business interests, whom they have consistently and zealously supported when in office, while the Democrats are allegedly sustained by a wide range of groups in civil society, including unions, teachers, low income people, and ethnic minorities. In this reading, support for the GOP should rise monotonically with median family income. Unfortunately, the reality is more complex, and less hopeful from the perspective of those concerned about the poor. Contrary to the stereotype, in fact statistically it was wealthier counties that were more likely to vote for Obama, while McCain garnered a majority in the poorest counties (Table 8.1). Thus, Obama, on average, won 58 percent of counties with above-average household incomes, including almost 63 percent of those with incomes over 120 percent of the national median, while McCain gained a slight majority of the vote in the poorest counties, i.e., those with incomes under 70 percent of the national median. In short, support for Obama rose steadily as median incomes rose, while they declined for McCain.

There are several forces at work that underpin this apparent discrepancy. Since the 1970s, the United States has been in the grip of a tremendous social transformation wrought by globalization, deindustrialization, stagnant incomes, rising inequality, growing demographic diversity, and wholesale class war launched by the ruling class under the banner of neoliberalism. Given its long history of individualism, anti-intellectualism, nationalist militarism, religiosity, anti-government rhetoric, obsession with a mythologized "free market," lack of social democracy, and weakened labor unions, no major political party effectively represents the interests of the working class, the poor, or the increasingly beleaguered middle class. Rather than a turn to

Table 8.1 Votes in 2010 Presidential Election by Household Income

Household Income as % of Average	Votes for Obama	% of Votes	Votes for McCain	% of Votes
> 120	7,512,019	62.8	4,450,295	37.2
100–119	8,721,561	55.1	7,118,178	44.9
90.0–99.9	8,556,514	57.8	6,258,803	42.2
80.0–89.9	15,958,416	55.1	13,021,362	44.9
70.0–79.9	14,570,760	51.2	13,869,966	48.8
< 69.9	14,159,650	48.2	15,216,153	51.8
Total	69,478,920	53.7	59,934,757	46.3

Source: Calculated by author.

working-class politics, mounting frustration at stagnant or declining standards of living, as well as deep seated fears about declining white hegemony, have driven countless millions of middle-income voters into the hands of conservatives, who preach a doctrine that substitutes liberals for the ruling class. The rise of conservatives since the 1970s, pioneered by Ronald Reagan and propelled by right-wing think tanks and conservative hegemony of the media (e.g., Fox News), has found widespread popular support among white Americans, particularly blue collar males, who often blame their economically precarious positions on the modest gains of marginalized groups such as women, immigrants, ethnic minorities, and gays. As a result, white Democratic voters began turning to the Republican Party en masse in the 1970s and 1980s, most particularly in the South (Aistrup 1996).

As Thomas Frank (2004) noted in his influential book *What's the Matter with Kansas*?, much of the popularity of American conservativism among the working class owes its existence to the propagation of a mythologized view of America as a uniquely blessed country increasingly overwhelmed by satanic force of moral deviance, as manifested by the rise of social indicators such as out-of-wedlock births, abortion ostensibly available on demand, multiculturalism, coddled criminals, and conservatives' favorite obsession, sexual permissiveness in various forms. In this reading, the roots of stagnant social mobility lay with the changes initiated during the 1960s, when evil began, an idea at the heart of many of the "cultural wars" fought over the last few decades. This notion of increasing liberal-promulgated decadence at home was accompanied by alarmist fears of declining US hegemony abroad. Neoconservatives have been highly successful at representing American politics in terms of a moral contest between good and evil, in which liberals, feminists, gays, immigrants, and environmentalists are equated with economic decline and moral decadence. Republicans have long popularized a mythologized view of a "liberal elite" intent of pushing a politically correct agenda down the throats of the innocent and young, whether through the corrupting values of the allegedly "liberal media" (including Hollywood) or the poisonous lectures of college professors. The real culprits, in this view, are those who have made modest gains allegedly at the expense of white males. Widespread ignorance and anti-intellectualism among a frighteningly gullible public contributed to the propagation of this worldview. Anti-intellectualism is nothing new in American political history (Hofstadter 1966), and is refreshed anew each generation by an educational system carefully designed to minimize any sense of criticality. The celebration of ignorance as a virtue among conservatives, for example, is evident in the growth of "tea party" movements following Obama's election, who soundly denounce government spending so long as it doesn't affect them.

The rhetoric of the radical neoconservative revolution admirably served the interests of neoliberalism, obfuscating the economic agenda of conservatives behind the rhetoric of family values, individualism, religiosity, and the ideology of the free market. In essence, the Republican Party has used a faux populism to disguise its deeply elitist politics, which are designed to assist the ruling class in extracting an ever-larger share of the country's surplus value. Thus, under the administrations of Ronald Reagan and George W. Bush, income inequality grew markedly, in part due to deregulation and tax cuts on unearned income, which is the primary source for the well-to-do. In this context, what voters expect and what they receive are often two different things. As Frank (2004, 7) puts it:

> *Vote* to stop abortion; *receive* a rollback in capital gains taxes. *Vote* to make our country strong again; *receive* deindustrialization;. *Vote* to screw those politically correct college professors; *receive* electricity deregulation. *Vote* to get government off our backs; *receive* conglomeration and monopoly everywhere from media to meat-packing. *Vote* to stand tall against terrorists; *receive* Social Security privatization. *Vote* to strike a blow against elitism; *receive* a social order in which wealth is more concentrated than ever before in our lifetimes, in which workers have been stripped of power and CEOs are rewarded in a manner beyond imagining.

Essentially, such a strategy allows Republican candidates such as McCain to claim to speak on behalf of the middle class without offending their big business constituency. Many middle class voters, especially whites, are led enthusiastically to vote against their own economic interests, leading to what Frank labels a "French Revolution in reverse." Veterans, for example, eagerly supported McCain in 2008 even as the GOP voted to cut their benefits. It is important to note that the neoliberal agenda that has blamed government and liberals as the source of the country's problems has, with admirable efficiency, made great strides in rolling back many of the social and policy gains made during the 1960s and the Great Society, is now targeting initiatives launched during the New Deal (e.g., Social Security), and appears eagerly intent on taking the country back to a nineteenth century form of laissez faire and social Darwinism.

Ethnicity/Race: The Fault Line of American Politics

The influence of race or ethnicity on American voting patterns has been the subject of considerable scrutiny (Leighley and Vedlitz 1999; Hutchings and Valentino 2004), although its effects are difficult to disentangle from those of class and other bases of identity. Although there are many nuances to this story, in essence, since the 1970s Republican supporters, especially the most vehement ones, tend to be disproportionately white, whereas Democrats have traditionally attracted support from ethnic minorities. This pattern reflects, in no small part, the success of the GOP in capitalizing on white resentment against minorities for social and economic gains (often exaggerated in the minds of conservatives) that ostensibly come at the expense of working class whites. The GOP's famous "Southern Strategy" initiated in the 1970s, for example, led to widespread white abandonment of the Democrats in the South and the entrenchment of GOP politicians in what was once solidly Democratic territory (Black and Black 2002). By diverting white resentment toward minorities, the fundamental role of race and ethnicity has been to obfuscate the role of class. As Huckfeldt and Kohfeld (1989, 12) note, "race served to splinter the Democratic coalition" because the federal policies initiated during the Civil Rights era provoked "[r]acial hostility, particularly on the part of lower-status whites." However, the mounting conservativism of American whites has, to some extent, been offset by the growing minority populations, now roughly 30 percent of the population, which tend Democratic. The rapid growth of Latinos, for example, who now form the country's largest ethnic minority, has problematized the traditional black-white binary (Segura and Rodriguez 2006).

Ethnicity also played an important role in the 2008 presidential election (Ansolabehere and Stewart 2009; Mas and Moretti 2009; Pasek et al. 2009; Philpot et al. 2009). Obama's racial composition obviously played a hugely important role. Widely touted as the first "African-American presidential candidate," Obama is in fact of mixed race (African father, white American mother), but in the racial discourses of American politics being of mixed race is tantamount to being equated with "black," a nomenclature reminiscent of the old racial stratification system found in the antebellum South. Race figured prominently in the heated Democratic primary contest between Obama and Hillary Clinton, who enjoyed more support among whites. Despite his attempts to appear "post-racially" transcendent on this issue, Obama suffered repeated and blatantly false conservative allegations that he was a Muslim, not born in the US, or was a closet Marxist, all of which reflected a deep racism on the part of a large chunk of the electorate. Obama's ethnicity, as well as his foreign-sounding name, helped Republican efforts to demonize otherness, albeit unsuccessfully, fueling latent white xenophobia and concerns about their declining proportion of the population. More subtly, even if they are not explicit racists, most whites exhibit a preference for white candidates (Bafumi and Herron 2009). Nonetheless, despite such obstacles, and due in part to widespread disgust with the Bush administration and a collapsing economy, Obama's campaign cobbled together a successful coalition of ethnic minorities and progressive whites, including Democrats, many independents, and even a few "Obamicans" (Republicans who voted for Obama).

While Obama's election was undoubtedly a historical achievement in the struggle for racial equality, the 2008 presidential election lent support to the notion that most backers of Democrat candidates are minorities and most Republican supporters are found in predominantly white regions. Table 8.2, for example, summarizes exit polls on the eve of the election, indicating that whites preferred McCain by a margin of 55 to 43 percent, whereas Obama enjoyed overwhelming popularity among blacks (95 percent), who were electrified by his candidacy, and solid majorities among Latinos and Asian-Americans. Given the numerical dominance of whites among the electorate, such a pattern would seem to indicate an inevitable victory for McCain. However, within the complex dynamics of the electoral college and its winner-take-all rule for amassing electoral votes, white support for McCain was largely rendered ineffectual, in part because it was so heavily concentrated in the South. In this sense, the Southern Strategy may have worked against Republicans. As Table 8.3 reveals, while minority-dominant

Table 8.2 Exit Polls of Voters by Ethnicity on Eve of 2008 Presidential Election

Ethnicity	% of the Electorate	% for Obama	% for McCain
White	74	43	55
Black	13	95	1
Latino	9	67	31
Asian	2	62	35

Source: *New York Times*. 2008. November 5. National Exit Polls Table. Available at: http://elections.nytimes.com/2008/results/president/national-exit-polls.html.

Table 8.3 Votes in 2010 Presidential Election by White Percent of County Population, 2008

% Population White	Votes for Obama	% of Votes	Votes for McCain	% of Votes
> 90	31,783,205	53.2	27,993,460	46.8
80.0–89.9	11,260,897	52.1	10,365,811	47.9
70.0–79.9	8,056,834	54.5	6,736,108	45.5
60.0–69.9	7,142,757	52.6	6,442,390	47.4
50.0–59.9	5,306,683	53.0	4,697,883	47.0
< 50.0	5,928,544	61.6	3,699,105	38.4
Total	69,478,920	53.7	59,934,757	46.3

Source: Calculated by author.

counties voted heavily for Obama, in majority-white counties Obama still amassed a slim majority of votes. In short, many whites did indeed support Obama, and there are important geographic patterns that are hidden by these national data.

Religion and American Presidential Politics

The role of organized religion in American politics is too important to ignore. From popular beliefs of the electorate and politicians to how issues are often discursively framed, religion saturates many political debates. From the country's founding, religion and politics were intertwined (Thompson 1988), as exemplified historically by the anti-slavery movement, agrarian socialism, Prohibition, and the civil rights struggle. In contrast to most of the industrialized world, such as Europe, in which secularism is more common than religiosity, the US exhibits significantly higher levels of religious adherence and participation (Hackett 2003). Religion's political influence reflects the unique circumstances of the United States – as claimed by the American exceptionalism thesis – including the peculiarly high degrees of religiosity found in the US.

A large political science literature has investigated how religion has played a critical role in shaping the political loyalties of vast swaths of the electorate (e.g., Green et al. 2002; Leege et al. 2002; Wald and Calhoun-Brown 2007). Religion enters electoral politics through a variety of channels, including how it shapes the ideological propensities of the electorate and candidates as well as the inclination to vote itself. For example, large majorities of Americans indicate they would not vote for a candidate "who did not believe in God' (Servin-Gonzalez and Torres-Reyna 1999). Religious attitudes are also important to politicians, including presidential candidates and members of Congress, who "are not empty religious vessels, but indeed bring deeply held religious conceptions to the world" (Hertzke 1988, 829). Often Congressional religious preferences and voting records reflect the relative degree of religiosity in their respective districts. For example, generally, significant local membership in conservative Protestant churches is inversely related to liberal Congressional voting records (Green and Guth 1991). The relations between theological and political ideology are complex, however, and should not be oversimplified. While some maintain that religion inculcates a passive acceptance of the prevailing social order, others point to the role of religion in progressive trends such as the civil rights movement.

The rise of politically conservative Protestants is perhaps the most visible recent face of this phenomenon. In the late twentieth century, evangelical conservative religious activists surfaced as a highly influential political force (Guth and Green 1991; Wilcox 1992). In 1976, for example, Jimmy Carter ran as a "born again" Christian, and religion was openly embraced by Ronald Reagan and Bill Clinton. In the 1980s, leaders of the Republican Party, recognizing the size, militancy and potential power of this electorate, began to court such voters aggressively, leading them to become a pillar in a highly successful political alliance with business interests. The Moral Majority and religious leaders such as Jerry Falwell and Pat Robertson became vocal and influential voices on behalf of conservative Protestantism in particular and the Republican Party more generally. Today, many GOP state political platforms affirm the status of the US as a Christian nation, label the separation of church and state a myth, demand abstinence instead of sex education, reject evolution, and otherwise mirror the agenda of religious conservatives (Phillips 2006). Thus, Heinemann (1998) observes, the predominant impact of contemporary religion in the US has been to inspire, motivate, and produce conservatives. As Jacobs (2006) notes, the marriage of political and religious conservatives occurred precisely at the historic moment in which American economic and political hegemony was contested internationally, a moment in which multiculturalism and the postmodern relativity of moral values achieved widespread popularity. Indeed, the rise of the Religious Right fueled not only a series of "culture wars," but also a significant gender and ethnic gap in which white males increasingly turned toward the Republican Party whereas women and ethnic minorities remained predominantly Democratic in their voting behavior. Issues that particularly animate conservative religious voters include prayer in the schools, reproductive rights and contraception, sex education, tolerance of homosexuality, the teaching of evolution, stem cell research, and a variety of other important topics often loosely lumped under the phrase "moral values." As a result, the "religion gap" between Republicans and Democrats has become one of the most striking and important dimensions of US politics, rivaling or exceeding the significance of class, race, or gender in determining local and national electoral turnout and outcomes.

Religion played an important role in the 2008 presidential election (Espinosa 2008). President George W. Bush, a born-again Methodist, personified conservative Protestantism, with numerous public references to the role of God in his personal and public life (Mansfield 2003; Aikman 2004). He was arguably the most visibly religious president in decades, perhaps in national history. However, given widespread voter revulsion at the Bush presidency, the effects of religion played out differently among different denominations. For example, counties that were heavily Catholic tended to favour Obama by wide margins (Table 8.4), in part due to his popularity among Hispanics, whereas conversely McCain won substantial victories in the counties in which Baptists comprised 30 percent or more of county adherents (Table 8.5).

Table 8.4 Votes in 2008 Presidential Election by County Percent of Total Adherents who are Catholic

% Adherents Catholic	Votes for Obama	% of Votes	Votes for McCain	% of Votes
> 70.0	3,447,288	61.1	2,196,309	38.9
60.0–69.9	5,675,259	58.0	4,116,250	42.0
50.0–59.9	10,973,988	59.4	7,487,087	40.6
40.0–49.9	7,494,705	56.5	5,778,437	43.5
30.0–39.9	8,348,764	55.6	6,674,328	44.4
< 30.0	33,538,917	49.9	33,682,344	50.1
Total	69,478,920	53.7	59,934,757	46.3

Source: Calculated by author.

Table 8.5 Votes in 2008 Presidential Election by County Percent of Total Adherents who are Baptist

% Adherents Baptist	Votes for Obama	% of Votes	Votes for McCain	% of Votes
≥ 70.0	941,804	46.6	1,078,921	53.4
60.0–69.9	2,788,220	43.8	3,574,970	56.2
50.0–59.9	3,355,889	42.3	4,574,288	57.7
40.0–49.9	2,851,024	43.8	3,650,743	56.2
30.0–39.9	4,815,032	48.8	5,048,211	51.2
< 30.0	54,726,951	56.6	42,007,624	43.4
Total	69,478,920	56.6	59,934,757	46.3

Source: Calculated by author.

Table 8.6 Simple Correlation Coefficients between Percent of Votes for Obama and McCain and Income, Ethnicity, and Religious Denomination

	% Voting for Obama	% Voting for McCain
Household Income	0.26	-0.25
% White	-0.31	0.28
% Black	0.27	-0.24
% Latino	0.06	-0.05
% Asian	0.25	-0.25
% Catholic	0.36	-0.37
% Baptist	-0.41	0.43
% Methodist	-0.24	0.25
% Pentacostal	-0.13	0.12
% Latter-day Saints	-0.15	0.13
% Jewish	0.32	-0.32
% Muslim	0.18	-0.17

The Independent Effects of Class, Race, and Religion: Empirical Assessments

The simple relationships between the variables discussed above and votes for Obama and McCain can initially be approximated using simple correlation analysis (Table 8.6; given the large sample size, 3,138 counties, all of these coefficients are statistically significant). These numbers confirm many of the observations made above. Thus, average household income was strongly and positively correlated with support for Obama (.26), and negatively so with votes for McCain (-.25). Percentage white was strongly associated with opposition to Obama (-.31) and support for McCain (.28) while percentage black and Asian-American exhibited exactly the reverse trend; Latinos were not strongly associated with either candidate. Among religious denominations, Catholics, Jews, and Muslims supported Obama and opposed McCain, while Baptists, Methodists, Pentacostalists, amd Latter-day Saints (Mormons) demonstrated the reverse pattern. Indeed, the correlations for the two candidates are largely mirror images of one another.

However, simple correlations disguise as much as they reveal. First, the effects of class, race, and religion on electoral outcomes are hopelessly intertwined with one another and empirically difficult to segregate. Second, the impacts of even one variable are likely to vary spatially: For example, household income may have very different effects on the West Coast compared to the South. For these reasons, this chapter analyzed votes for Obama using a relatively recent methodology, geographically weighted regression (GWR), which allows not only independent effects of variables to be specified, but also allows for spatial variations in the influence of one variable (Brundson et al. 1998; Fotheringham et al. 2002).

Standard regression analysis assume spatial stationarity, i.e., that the relationship under analysis is constant over space. In contrast, GWR permits relationships to vary spatially within a study region (or spatial non-stationarity). Each observation receives a weight that is a function of its proximity to the center of a given window (in this case, 20 counties). While standard regression analysis finds correlations between two variables across an entire data set, GWR yields correlations across only a geographic subset of the entire data set and then estimates how its statistical significance varies across the region under study.

Including variables derived from the discussion above (but limiting religious denominations to only two), the multivariate linear GWR model utilized in this analysis is

$$p_{oi} = \beta_i + \beta_{hi} + \beta_{wi} + \beta_{bi} + \beta_{li} + \beta_{ai} + \beta_{ci} + \beta_{bi} + \varepsilon,$$

where

p_{oi} = % voting for Obama in each county i;
β_i = constant
β_{hi} = county household income as % of national median;
β_{wi} = percent white population in county i;
β_{bi} = percent black population in county i;
β_{li} = percent Latino population in county i;
β_{ai} = percent Asian population in county i;
β_{ci} = % of adherents who are Catholic in county i;
β_{bi} = % of adherents who are Baptist in county i;
ε = error term.[1]

Such an approach allows for a view of the geographically uneven independent effects of each variable controlling for the influence of the others. The results of the analysis for each variable are summarized in maps of the partial regression coefficients, and, when necessary, the results of a simple GWR bivariate model without controls as well as the results of the broader model with controls.

GWR Class Effects

Although it is an imperfect measure of class, income (in this case, county average income as a percent of the national median) is an important determinant of voter preferences. The spatial distribution of household incomes (Figure 8.3) reflect the uneven development that has long formed a centerpiece of the social and economic geography of the United States, with the highest levels found in the northeastern seaboard, the West Coast, much of the old Manufacturing Belt, and metropolitan

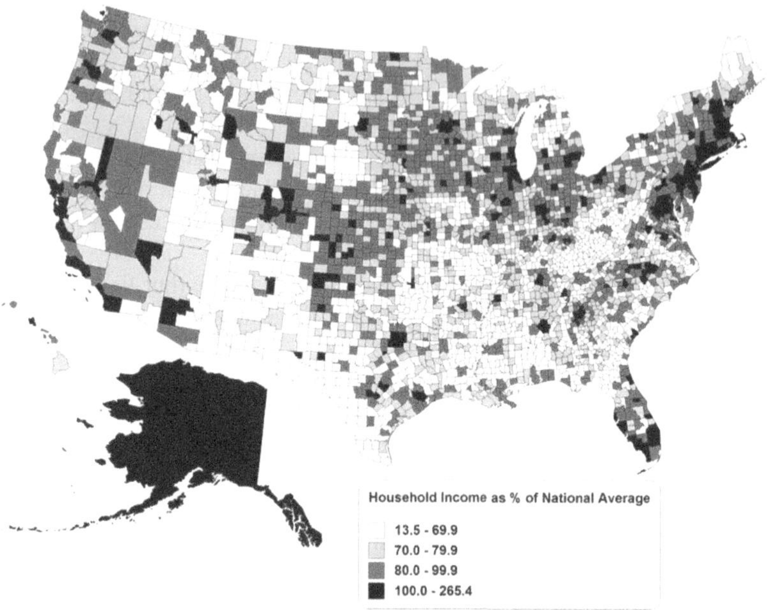

Figure 8.3 Household Income as a Percent of the National Average

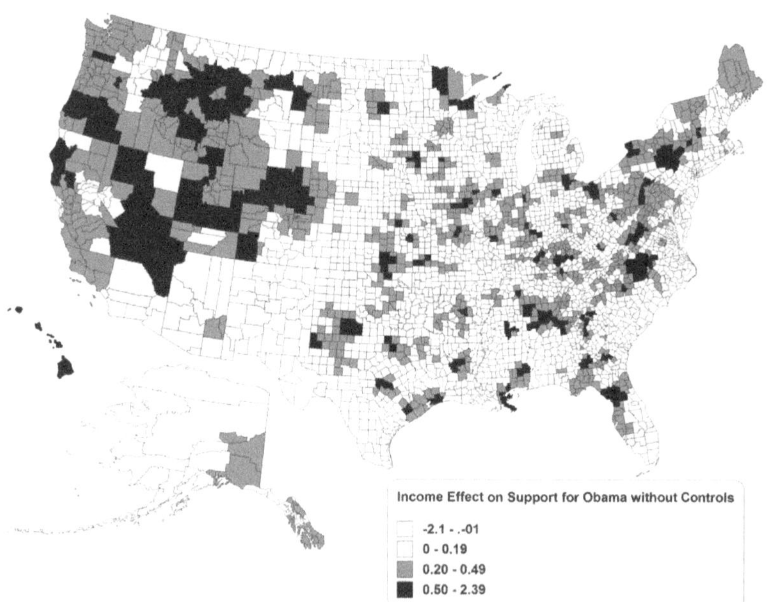

Figure 8.4 Results of Bivariate GWR of Support for Obama and Average Household Income without Controls for Ethnicity and Religion

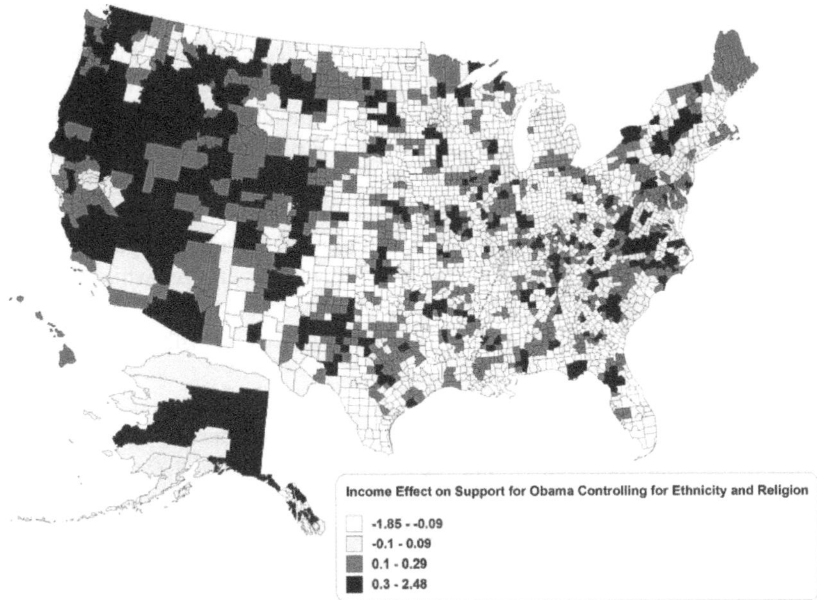

Figure 8.5 Independent Effects of Household Income on Support for Obama, Controlling for Ethnicity and Religion

areas generally, and lower levels distributed across rural areas, particularly in the South. A simple bivariate GWR analysis of income and support for Obama yields surprising results (Figure 8.4): higher income households demonstrated the most support for him in traditionally conservative regions such as the intermontane West, notably Utah and Idaho, the West Coast, and pockets widely distributed throughout the central and eastern thirds of the country. However, the bivariate relationship may be heavily affected by other variables. Once the controls for ethnicity and religion are introduced, the GWR nonetheless confirms that Obama's support tended to rise steadily with household income in the West more than most parts of the country (Figure 8.5). In contrast to their counterparts elsewhere in the country, it would seem that upper income Western voters are relatively more Democratic in their inclinations, even in regions dominated by conservative religious denominations such as the Church of Latter-day Saints (Mormons). Such a result may reflect the particular cultural heritage of the Western United States, in which ideologies such as individualism and the cult of entrepreneurship are particularly strong (Etulain 1980).

GWR Ethnicity Effects

Although the voting impacts of ethnicity have been well demonstrated among ethnic groups, very little insight has been shed on their effects within a given ethnic group. GWR allows for such an analysis. Whites, who comprised roughly three-quarters of the electorate, demonstrated spatially uneven support for Obama. Thus, without the controls of income or religion, white support was highest in northern California and the Pacific Northwest (including southern Idaho!) and parts of Virginia near Washington, DC (Figure 8.6). Once the effects of income and religion are removed, however, this pattern changes significantly, with white support highest in the West Coast and Nevada, the Mississippi Delta, Colorado, southern New England, and the greater Washington, DC region (Figure 8.7). Such a pattern points to two possible forces contributing to a liberalized white electorate: the presence of a large minority population and local economies centered either on the public sector or professionalized producer services, as with Florida's (2004) celebrated "creative class."

Black support for Obama, which was almost universal, also demonstrated a geography in its own right. Without controls for income and religion, blacks voted for Obama most heavily outside of the South (Figure 8.8), particularly in the Great Plains and Rocky Mountains. Similarly, Latino support for Obama was heaviest outside of the Southwest, notably in parts of the South with large African-American populations, Illinois and Wisconsin, and New England (Figure 8.9). Such geographies, intimately tied up with the politics of identity, hint that in both cases minority support was accentuated in regions where they were less common numerically.

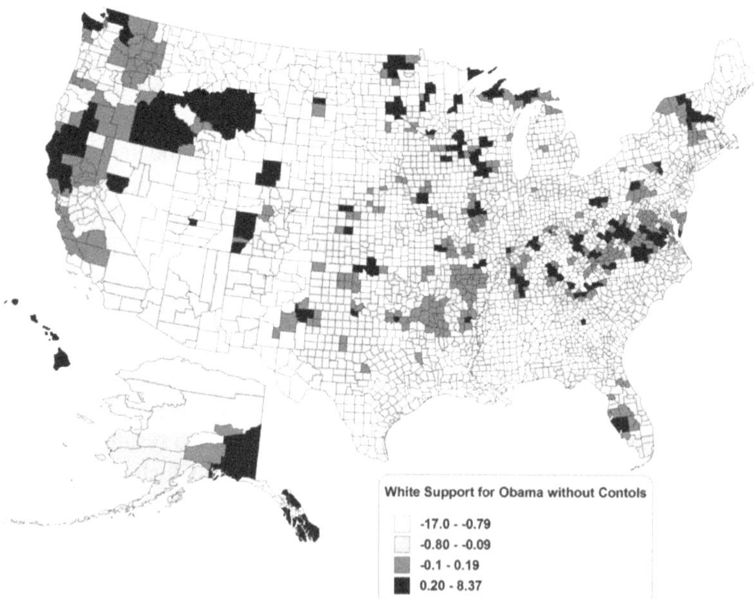

Figure 8.6 Results of Bivariate GWR of Support for Obama and White % of County Population without Controls for Income and Religion

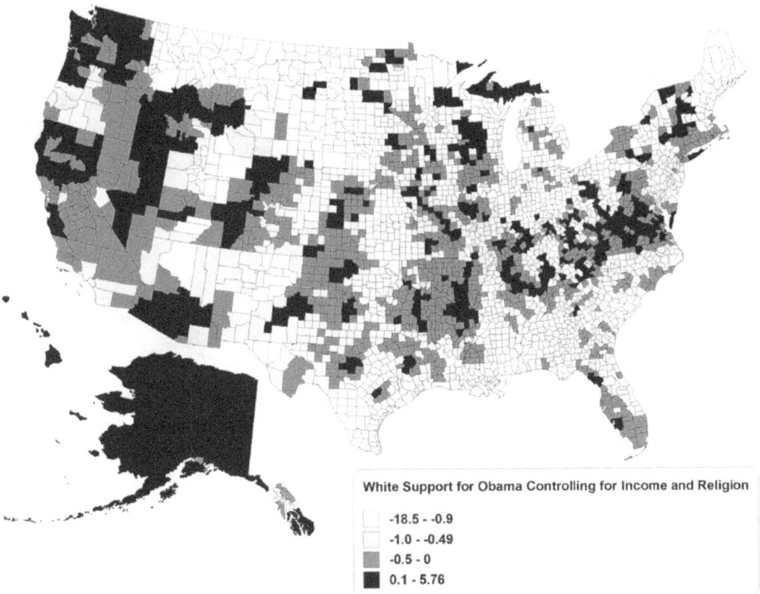

Figure 8.7 Independent Effects of White Support for Obama, Controlling for Other Ethnicities and Religion

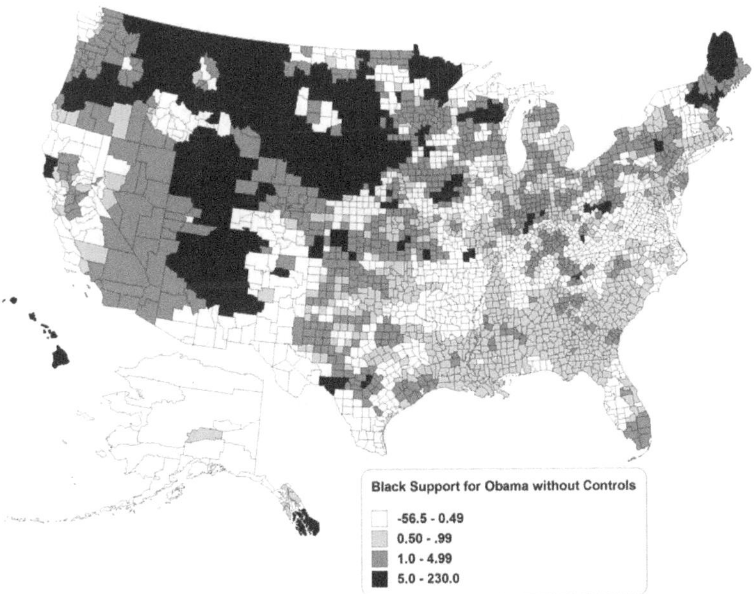

Figure 8.8 Results of Bivariate GWR of Support for Obama and Black % of County Population without Controls for Income and Religion

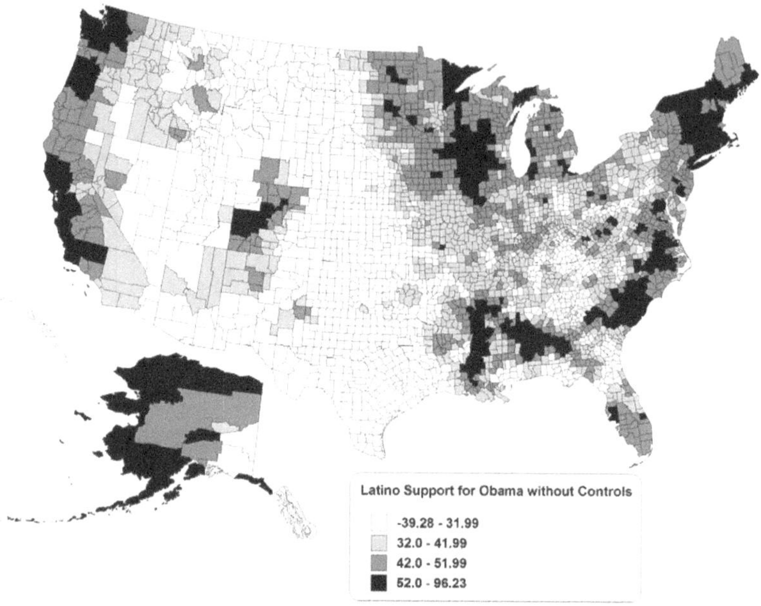

Figure 8.9 Results of Bivariate GWR of Support for Obama and Latino % of County Population without Controls for Income and Religion

GWR Religion Effects

Data on the distribution of religious denominations by county in 2000 were obtained from the census published by the Glenmary Research Center (2002). While the Glenmary data are widely recognized as the best available, they are not without flaws. Some denominations such as the Jehovah's Witnesses refused to participate. Also, the Glenmary database does not identify the religion of those who live in one county and worship in another. Because the data on religious adherents is geolocated by place of worship (which may be in a neighboring county) rather than place of residence, as in the case of voting data, there is the distinct possibility of a spatial mismatch between religious affiliation and voting patterns, although for purposes of studying the dynamics of the electoral college, which is based on states, not counties, this issue is irrelevant.

Catholics, for example, who comprise 37 percent of the electorate, form significant majorities in most of the Northeast, Wisconsin, and much of the Southwest, particularly California, New Mexico, and southern Texas (Figure 8.10). All of these regions voted for Obama in the general election. Catholics are an ethnically and politically diverse lot, of course, ranging from conservative Cuban-American voters in southern Florida to more liberally minded Puerto Ricans in Florida and New York, Italians and Irish in the Northeast, and Mexican-Americans

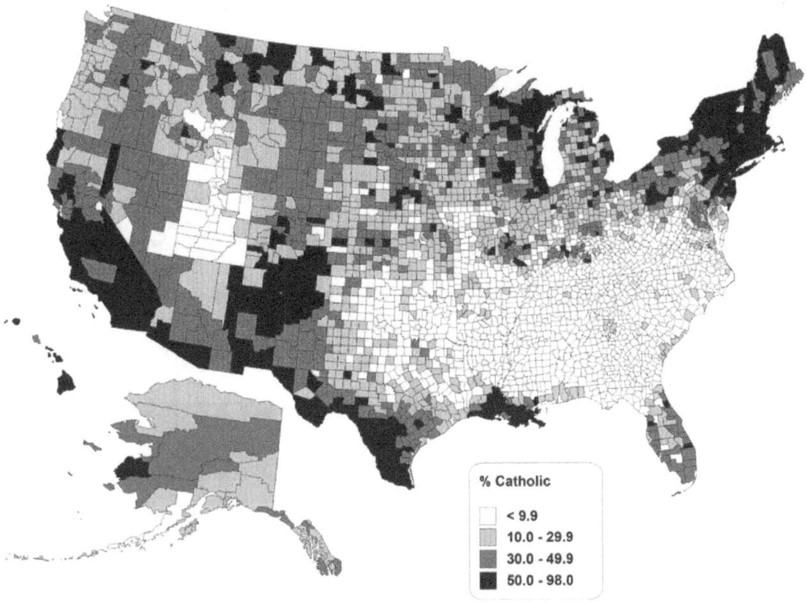

Figure 8.10 Catholics as % of Total Adherents

in the Southwest. GWR reveals that once the effects of income and ethnicity are removed statistically, Catholic support for Obama consisted of a hodgepodge of pockets of greater and lesser enthusiasm (Figure 8.11). Catholics in the South tended to vote him in greater proportions than elsewhere, although Catholics in southern New Mexico and Utah likewise were also supportive. Surprisingly, Catholic support in strongholds such as New England and the northern Midwest was relatively tepid.

Protestant denominations are quite another story. Baptists (14.3 percent of voters), who comprise the largest Protestant group, generally voted heavily in favor the Republican John McCain. Concentrated heavily in the South, particularly interior counties removed from the coasts, Southern Baptists were located in states that all voted against Obama by large margins (Figure 8.12), although Florida, North Carolina, and Virginia did narrowly vote Democratic. Baptist support for Obama, such as it was lay for the most part outside of the South (Figure 8.13), including islands of support around the New York metropolitan region, Chicago, the northern Great Plains, the borderlands of Texas and New Mexico with Mexico, and the Pacific Northwest. Such a pattern reflects, perhaps, the willingness to consider candidates not actively supported by one's denomination when living as a minority outside of the region of denominational hegemony.

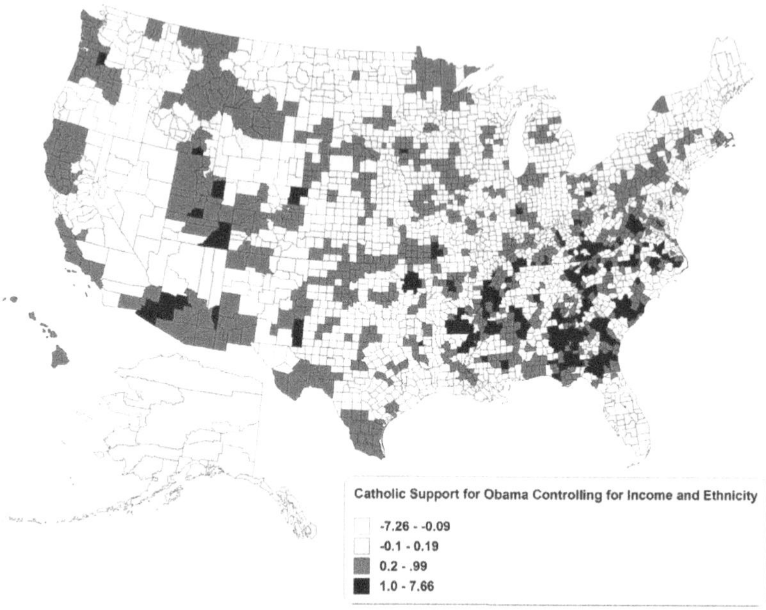

Figure 8.11 Independent Effects of Catholic Support for Obama, Controlling for Income and Ethnicity

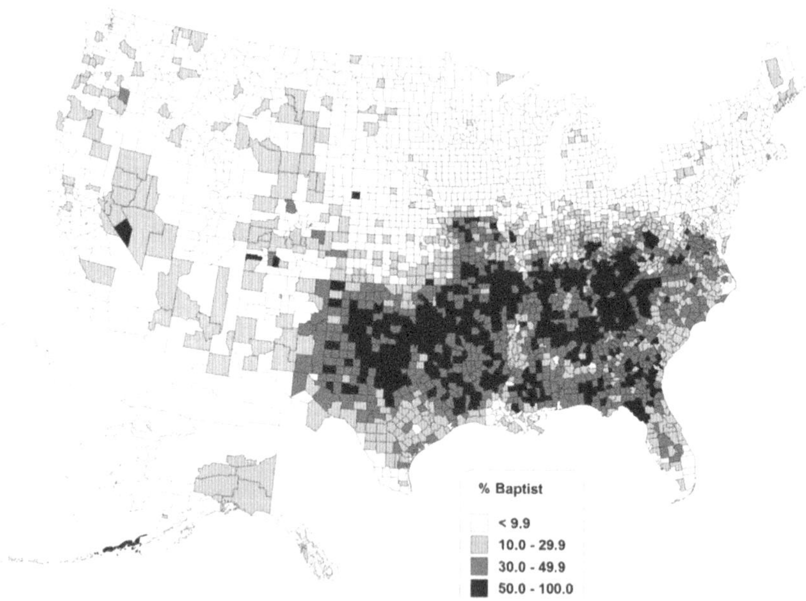

Figure 8.12 Baptists as % of Total Adherents

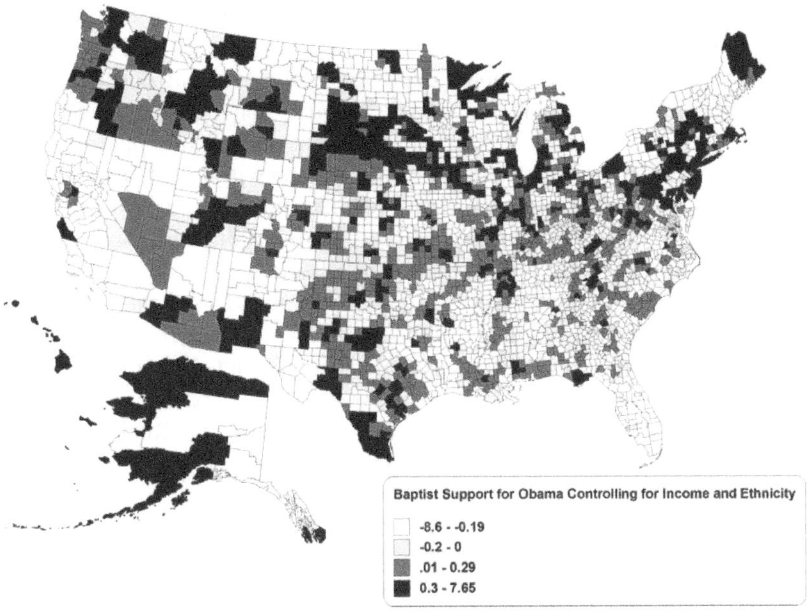

Figure 8.13 Independent Effects of Baptist Support for Obama, Controlling for Income and Ethnicity

Concluding Thoughts

The presidential contest of 2008 was a watershed moment in several respects, including a rejection of the policies of the George W. Bush administration and the election of the country's first African-American candidate. Obama's victory reflected a constellation of voters of many kinds with varying reasons. In this race, as always, class, ethnicity, and religion all had profound effects on the American electoral landscape. (No doubt so too did other markers such as age, gender, and sexuality). Contrary to much received opinion, among income groups, the Democrats' greatest strengths lay among upper income voters rather than poorer households, who generally supported McCain. Whites, at least outside of the South, tended to be split between the two parties, whereas blacks, Latinos, and Asians tended to support Obama whole heartedly. Using the two largest denominations as a marker of religious influence, Catholics tended to vote for Obama in large proportions, whereas Baptists were strongly opposed to his candidacy.

What is critical to note here is that the effects of these variables are not spatially uniform: contrary to the aspatial wonderland in which many economists and political scientists operate, there are enormous geographical variations in the intensity of the relationship of each category. Geographically weighted regression offers a valuable means with which to interrogate such patterns in a statistically rigorous manner.

The results of the GWR model deployed here hint at one consistent, albeit surprising, finding: support for Obama within each economic, ethnic, and religious group tended to be highest in regions in which supporters were a relative minority. In most cases, support for Obama within each income, ethnic, or religious group was greatest when members of that group were relatively uncommon. Whites, the majority in most regions, did not conform to this generalization. However, black support for Obama was greatest outside of the South, and Latino support highest in regions removed from the Southwest or Florida. Some Baptists did indeed support Obama, but largely in areas outside of the South. Higher levels of household income tended to be more closely associated with votes for Obama in the Western US (parts of which are quite wealthy, belying the observation made above), including the Rocky Mountains, Nevada, and California. At the least, these results indicate the need to avoid simplistic, aspatial generalizations that essentialize groups and their political views.

More deeply, such a trend points to the politics of marginality that played an important role in electing Obama. If identity is largely constructed and maintained through difference (Sibley 1995; Natter and Jones 1997; Valentine 2008), then voters belonging to particular income, ethnic, or religious categories who live in regions in which they find themselves minorities may well have been more sympathetic to a candidate who was a member of an ethnic minority himself. Perhaps supporters of Obama who were not surrounded by others of similar wealth, culture, and beliefs were more willing to elect a candidate whose own background was so resoundingly non-traditional. In short, a geographical analysis indicates that the lines of difference so important not only to identity but to political attitudes are highly susceptible to local and regional variations.

Acknowledgement

The author thanks Aaron Gilbreath for his assistance with the GWR.

References

Aikman, D. 2004. *A Man of Faith: The Spiritual Journey of George W. Bush*. New York: W Publishing Group.

Aistrup, J. 1996. *The Southern Strategy Revisited: Republican Top-Down Advancement in the South*. Lexington, KY: University of Kentucky Press.

Ansolabehere, S. and C. Stewart. 2009. "State of the Nation: How Post-Racial was Obama's Victory?" *Boston Review* 34(1): January/February.

Bafumi, and Herron. 2009. “Prejudice, Black Threat, and Racist Voters in the 2008 Presidential Election.” *Journal of Political Marketing* 8(4): 334–48.

Black, E. and M. Black. 2002. *The Rise of Southern Republicans*. Cambridge, MA: Harvard University Press.

Brunsdon, C., S. Fotheringham, and M. Charlton. 1998. “Geographically Weighted Regression.” *Journal of the Royal Statistical Society* 47(3): 431–43.

Espinosa, G. 2008. *Religion, Race, and the American Presidency*. Lanham, MD: Rowman and Littlefield.

Fowler, R., A. Hertzke, L. Olson, and K. den Dulk. 2004. *Religion and Politics in America: Faith, Culture, and Strategic Choices*. Boulder, CO: Westview Press.

Fotheringham, A., C. Brunsdon, and M. Charlton, M. 2002. *Geographically Weighted Regression: The Analysis of Spatially Varying Relationships*. Chichester: Wiley.

Etulain, R. 1980. “Frontier, Region and Myth: Changing Interpretations of Western American Culture.” *Journal of American Culture* 3(2): 268–84.

Florida, R. 2004. *The Rise of the Creative Class*. New York: Basic Books.

Frank, T. 2004. *What's the Matter with Kansas? How Conservatives Won the Heart of America*. New York: Holt.

Glenmary Research Center. 2002. *Religious Congregations and Membership in the United States*. Atlanta: Glenmary Research Center.

Green, D., B. Palmquist, and E. Schickler. 2002. *Partisan Hearts and Minds*. New Haven, CT: Yale University Press.

Green, J. and J. Guth. 1991. “Religion, Representatives, and Roll Calls.” *Legislative Studies Quarterly* 16: 571–84.

Green, J., M. Rozell, and C. Wilcox (eds) 2006. *The Values Campaign? The Christian Right and the 2004 Elections*. Washington, DC: Georgetown University Press.

Guth, J. and J. Green (eds) 1991. *The Bible and the Ballot Box: Religion and Politics in the 1988 Election*. Boulder, CO: Westview Press.

Gutterman, D. 2005. *Prophetic Politics: Christian Social Movements and American Democracy*. Ithaca, NY: Cornell University Press.

Gutterman, D. and A. Murphy (eds) 2006. *Religion, Politics, and American Identity: New Directions, New Controversies*. Lanham, MD: Lexington Books.

Hackett, D. 2003. *Religion and American Culture*. London: Routledge.

Heilemann, J. and M. Halperin. 2010. *Game Change: Obama and the Clintons, McCain and Palin, and the Race of a Lifetime*. New York: Harper.

Heineman, K. 1998. *God is a Conservative: Religion, Politics, and Morality in Contemporary America*. New York: New York University Press.

Hertzke, A. 1988. “American Religion and Politics: A Review Essay.” *The Western Political Quarterly* 41: 825–38.

Hofstadter, R. 1966. *Anti-intellectualism in American Life*. New York: Vintage.

Huckfeldt, R., and C. Kohfeld. 1989. *Race and the Decline of Class in American Politics*. Urbana, IL: University of Illinois Press.

Hutchings, V. and N. Valentino. 2004. "The Centrality of Race in American Politics." *Annual Review of Political Science* 7: 383–408.
Jacobs, A. 2006. "The New Right, Fundamentalism, and Nationalism in Postmodern America: A Marriage of Heat and Passion." *Social Compass* 53: 357–66.
Jelen, T. 2006. "Religion and Politics in the United States: Persistence, Limitations, and the Prophetic Voice." *Social Compass* 53: 329–43.
Kohut, A., J. Green, S. Keeter, and R. Toth. 2000. *The Diminishing Divide: Religion's Changing Role in American Politics*. Washington, DC: Brookings Institute.
Leege, D., K. Wald, B. Krueger, and P. Mueller. 2002. *The Politics of Cultural Differences: Social Change and Voter Mobilization Strategies in the Post-New Deal Period*. Princeton, NJ: Princeton University Press.
Leighley, J. and A. Vedlitz. 1999. "Race, ethnicity, and political participation: Competing Models and Contrasting Explanations." *Journal of Politics* 61(4): 1092–114.
Mansfield, S. 2003. *The Faith of George W. Bush*. New York: Charisma House.
Mas, A. and E. Moretti. 2009. "Racial Bias in the 2008 Election." *American Economic Review* 99(2): 323–9.
Natter, W., and J.P. Jones. 1997. "Identity, Space, and Other Uncertainties." In G. Benko and U. Strohmayer (eds) *Space and Social Theory: Geographical Interpretations of Postmodernism*. Oxford: Blackwell, 141–61.
Olson, L. 2002. "Toward a Contextual Appreciation of Religion and Politics." In E. Blumhofer (ed.) *Religion, Politics, and the American Experience: Reflections on Religion and American Public Life*. Tuscaloosa: University of Alabama Press.
Pasek, J., A. Tahk, Y. Lelkes, J. Krosnick, B. Payne, O. Akhtar, and T. Tompson. 2009. "Determinants of Turnout and Candidate Choice in the 2008 US Presidential Election: Illuminating the Impact of Racial Prejudice and Other Considerations." *Public Opinion Quarterly* 73(5): 943–94.
Philpot, T., D. Shaw, and E. McGowen. 2009. "Winning the Race: Black Voter Turnout in the 2008 Presidential Election." *Public Opinion Quarterly* 73(5): 995–1022.
Phillips, K. 2006. *American Theocracy*. New York: Viking.
Segura, G. and H. Rodriguez. 2006. "Comparative Ethnic Politics in the United States: Beyond Black and White." *Annual Review of Political Science* 9: 375–95
Servin-Gonzalez, M. and O. Torres-Reyna. 1999. "Trends: Religion and Politics." *The Public Opinion Quarterly* 63: 592–621.
Sibley, D. 1995. *Geographies of Exclusion*. London: Routledge.
Thompson, K. 1988. "The Religious Transformation of Politics and the Political Transformation of Religion." *The Review of Politics* 50: 545–60.
Valentine, G. 2008. "Living with Difference: Reflections on Geographies of Encounter." *Progress in Human Geography* 32(3): 323–37.

Wald, K. and A. Calhoun-Brown. 2007. *Religion and Politics in the United States*. 5th edition. Lanham, MD: Rowman and Littlefield.

Wilcox, C. 1992. *God's Warriors: The Christian Right in Twentieth Century America*. Baltimore, MD: Johns Hopkins University Press.

Chapter 9
The Emerging Geography of How Americans Cast their Ballots

Toby Moore[1]

The first decade of the 2000s witnessed rapid change in how elections are run in the United States. Spurred on by the problems exposed by the Florida recount process after the 2000 presidential election, lawmakers, advocates and the public pressed for reform in a wide range of voting laws and practices. *Bush v. Gore* triggered a renewed sense that American election administration had become moribund and lax, in need of significant government investment and at least some measure of increased oversight and scrutiny. An initial focus was on voting technology, since the notorious "hanging chads" were the most visible sign of election failure in 2000, but other issues, from voter identification to voter registration to early voting, would also see new dynamism.

The landscape was actually already in some flux when *Bush v. Gore* hit, as a series of traditional concerns was prompting experimentation in the states and at the local level. Participation rates had long been a focus of political scientists and progressive reformers, who saw them as a measure of the health of American democracy, and new methods were adopted to make voting easier. Unsettled questions over the fairness of American elections, particularly to ethnic and racial minorities, and the continued salience of the Voting Rights Act of 1965 sustained efforts by the federal government to intervene in what is traditionally reserved for the states and local governments. Allegations of electoral fraud persisted, hotly contested by the political parties and the reform community. Electronic voting machines had already begun to replace paper ballots in some jurisdictions before Florida contemplated its hanging chads. The spark from the Florida recount process, therefore, hit dry tinder.

Reformers, however, confronted a country that stubbornly guards local prerogatives in determining how elections are run. The US famously lacks a constitutional right to vote, and the task of administering elections is left to the states, which in turn have historically devolved much responsibility to local jurisdictions. Filtered through the thousands of local election jurisdictions, then, the extreme federalism of US election administration was virtually guaranteed to

1 The author is a project director for the non-partisan, non-profit research institute RTI International. Generous support for this chapter was provided by RTI through a professional development award.

produce strong geographical variability as America grappled with how to improve its election methods.

This chapter describes the emerging spatial patterns of how Americans cast their ballots. It is "emerging" both because different changes are taking occurring in different places, and because we are only now collecting the data that makes seeing these patterns possible. This variance across space and between states and regions, it is argued, is of increasing importance to American democracy and the political system; the divergence appears to be growing, with potentially far-reaching effects. Investigation of this variation at the national level has only recently become possible due to newly-available data sources, including the federally-sponsored survey of election jurisdictions that forms the basis for this chapter.

While electoral geography has long been of interest to geographers, it has focused largely on the choices voters make or on the way electoral district boundaries are drawn. Little work has been done on the myriad ways that modern voters go about casting their ballots. Election administration has not generally been part of political geography, despite the rich geographic data available, to the detriment both of the discipline and American elections. Drawing largely from the results of the 2008 Election Administration and Voting Survey, sponsored by the US Election Assistance Commission and directed by the author, this chapter aims to lay the groundwork for a better engagement between geography and the question of how elections are run in the United States. Finally, the chapter discusses possible future directions for research, at scales from the local to the international, and ties the field to larger questions of democracy and equity.[2]

Elections in Electoral Geography

Traditionally, electoral geography has been divided into three branches (Taylor and Johnston, 1979). The first is the pattern of voters' choices, aggregated to a certain spatial scale and analyzed through mapping and multivariate analyses. Often socio-economic or other variables are included in the analysis in the search for understanding the spatial patterns. The second, "geographical influences on elections", was the subject of considerable debate on so-called "context effects" and their importance and value as a subject of inquiry (Agnew, 1996; King, 1996). This branch saw geographers defend the impact of local circumstances on voter choices, with mixed success. Finally, the "geography of representation" is focused on the drawing of district boundaries and the biases that develop from that process.

2 Unless stated otherwise, all the data used in this article are from the EAC survey and publicly available through the EAC's website at www.eac.gov. Summary tables with statewide aggregations are available in the reports based on the survey issued by the EAC and also available at the same Web address.

This has been a particular interest in the United States, where little is done to constrain partisan gerrymandering, and where racial gerrymandering (for both the benefit and harm of minority voters) has a long tradition, but has proved important in other countries as well. Border delineation is an important part of the technical assistance provided to emerging democracies, who must often build constituencies from scratch.

Within these three branches of electoral geography, little has been written about the geography of the myriad ways in which people differ in the way they cast their ballots. The "geography of elections," to date, has referred almost exclusively either the spatial variation in the choices that voters make between candidates, parties or issues, or the division of space into electoral districts. A "…geography of election administration", by contrast, would focus on the way laws and procedures impact the molding of the electorate and their role in promoting or restricting democratic outcomes, and the variance of such laws and procedures across the thousands of election jurisdictions in the United States. A broad array of topics can be studied in such a geography of election administration; the intention here is to suggest some topics that other geographers might find of interest for their own work, as well as providing a high-level description of some of the ways election laws and practices are diverging across the country. The topics discussed below, as a starting point for this geography of election administration, include:

- when voters cast their ballots (i.e., on election day or before, or even afterwards);
- where they cast their ballots (i.e., polling places or remotely)
- how they cast their ballots (i.e., through the mail or in person); and,
- what barriers they face (i.e., ID laws, felony disenfranchisement or registration requirements);
- what technology they use to record their choices (i.e., electronic voting machines or paper ballots).

Such issues are of growing interest to political scientists, legal scholars and election professionals. Much of the literature they have produced focuses on the impact of specific election reforms either on participation as a whole, or on the participation of specific parts of the electorate. For example, the move by some states during the 2000s to impose new identification requirements on voters (discussed below) slowly gave rise to a small body of scholarly writings over the second half of the decade. Not surprisingly, it lacked consensus and mirrored the sharp political divide over the issue, but nonetheless was able to provide policy makers with some of the basic research needed to consider and debate ID laws. Much of the work done on the impact of voter ID laws was done by scholars or advocates with an interest in one side of the debate or the other, an indication of the shortage in election studies of independent social science.

The 2008 Election Administration and Voting Survey

Under the mandate given it by the 2002 Help America Vote Act (HAVA), the US Election Assistance Commission began collecting election data following the 2004 election. Initially called the Election Day Survey, and later the Election Administration and Voting Survey (EAVS), the project is somewhat misnamed; it is less a survey than an accounting process. Across a number of election administration topics, the EAVS attempts to collect data at the local level, generally the county or parish, but in some instances the town or township. The survey is administered to the individual states, territories and the District of Columbia, which either generate responses from central databases or collect the data from their own sub-jurisdictions. Some of the data requested is required by federal law (to measure compliance with the federal National Voter Registration Act (NVRA) and the Uniformed and Overseas Citizen Absentee Voting Act (UOCAVA)), but much of the data requested is voluntary on the part of the states and local governments. The data collects forms the basis for a series of reports issued by the EAC.

The EAC's work coincided with a push by academics and reformers to develop better data collection on election administration, and took place during what the journalist James Fallows has called "the decade of big data" across a variety of public policy fields from education to health to finance (Fallows, 2010). The Pew Charitable Trust's Make Voting Work project promoted a "Data for Democracy" effort, hosting conferences and issuing reports aimed at increasing the quality and quantity of election data available. A group of scholars, mostly law professors, argued throughout the decade that a reliance on empirical data on election administration was a way to overcome partisan and ideological divisions and avoid the traditional reliance on anecdotal evidence (for example, Tokaji, 2005; Overton, 2006; Gerken 2009).

Starting with a limited effort in 2004, data collection has improved with each iteration, netting higher response rates on most data elements and providing greater levels of disaggregated data. The 2008 EAC survey was the most complete data collection yet. Its baseline number of jurisdictions (including states or territories that only submitted aggregate responses; counties or parishes; and townships in some states) was 4,517, a 45 percent increase from the 3,123 jurisdictions covered by the 2006 survey. (The number of jurisdictions varies from year to year because some states aggregate their township data to the county level, or submit only statewide totals. Numbers of responses to individual items varied considerably). Data was collected from each of the 50 states, the District of Columbia, Guam, Puerto Rico, the Virgin Islands and American Samoa. While still incomplete, the 2008 Election Administration and Voting Survey (EAVS) suggests that future data collections will come yet closer to providing a complete picture of how Americans cast their ballots.

The 2008 survey was composed of two parts[3]: a qualitative Statutory Overview, which sought to gather data on the laws, definitions and procedures used by the states, and the main, quantitative survey of 52 questions (many questions had multiple parts). The main questionnaire included questions on voter registration; registration and voting by overseas civilians and uniformed service members; polling place operations; voting technology; poll workers; and other election administration topics.

Data was collected through the state or territorial election offices. Data collection practices and capacity vary significantly across the country. Many states maintain central voter registration and voter history databases that can be readily queried for most of the information requested. In other states, little data is kept at the state level, and it instead must be collected individually from each local jurisdiction. Many of these jurisdictions are small and lack the staff or motivation necessary to complete the survey. In addition, states vary in the ways they define and code important election terms (such as absentee ballots). For these reasons the level of response to individual questions on the survey varies. For some items, data was widely available; for others, there are significant holes. Finally, states differ in their willingness to cooperate with the EAC on the survey; some states and local officials feel the survey unduly burdens their offices and constitute an unfunded mandate on the part of the federal government.

Spatial Variations of Elections in the United States

Results from the 2008 EAVS show the marked differences between states and regions in the United States in how voters cast their ballots and other aspects of election administration. While this chapter focuses on state- and regional-differences, the county-level data available from the EAC can be readily mined to highlight local differences as well. In addition, the EAVS supplies data on topics beyond those covered below.

Convenience Voting

The traditional method of voting in the United States has two main components. The first, and most common, is voting in-person, on election day, usually at a polling place near one's place of residence. The second component, discussed below, entails voting through an absentee ballot.

---

3 An additional data collection effort related to the survey was the collection of precinct-level data from five states, as part of a $10 million Congressional program. This data, not included in this analysis, is also available at the EAC's website. The five states participating were Illinois, Ohio, Wisconsin, Pennsylvania and Minnesota.

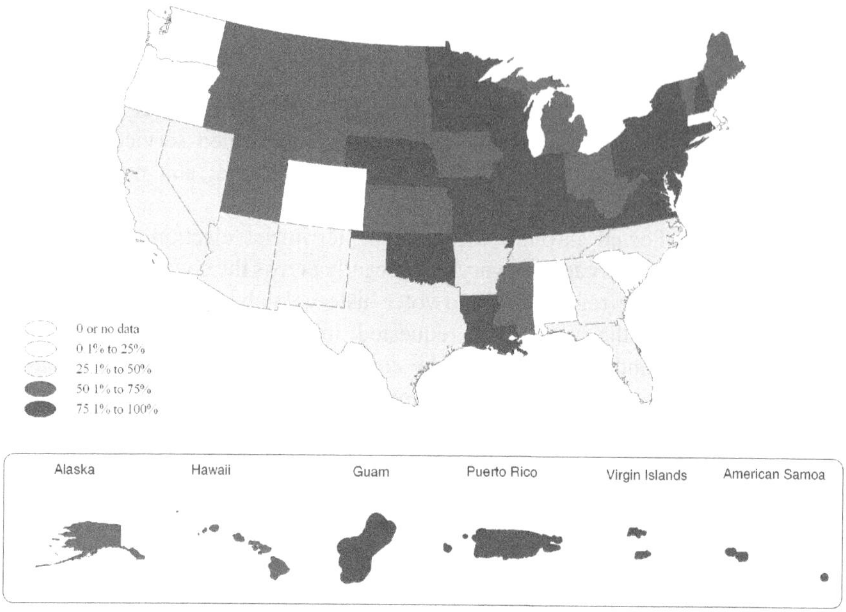

Figure 9.1 Percentage of Votes Cast In-person

Alongside this model have arisen a series of mechanisms intended to make it easier for people to cast their ballots, thus saving voters time and aggravation and encouraging participation. The actual impact on turnout of such measures is subject to debate; the literature at present suggests that the impact of most so-called convenience voting practices is minimal, and in some cases may even drive down overall turnout. States continue to experiment with convenience voting mechanisms, however, in part to save money through reduced administrative costs (Gronke, Galanes-Rosenbaum, Miller and Toffey, 2008).

The most effective way of measuring convenience voting is to look at the rates of traditional voting, since convenience voting comes in many forms. In 2008, states reported that 60.2 percent of voters cast their ballots in-person on election day (see Figure 9.1).[4] In-person voting on election day remains strongest in the Northeast and upper Midwest.

4 Not all states provided breakdowns of how their voters cast ballots; thus, another 6.1 percent of the ballots in the EAC survey were "not categorized". Many of these were undoubtedly in-person voters. Even within traditional balloting systems, some states and local jurisdictions have experimented with relaxing the geographic constraints imposed by polling places. "Vote centers", in which any registered voter in the jurisdiction can cast a ballot, have been implemented in some places.

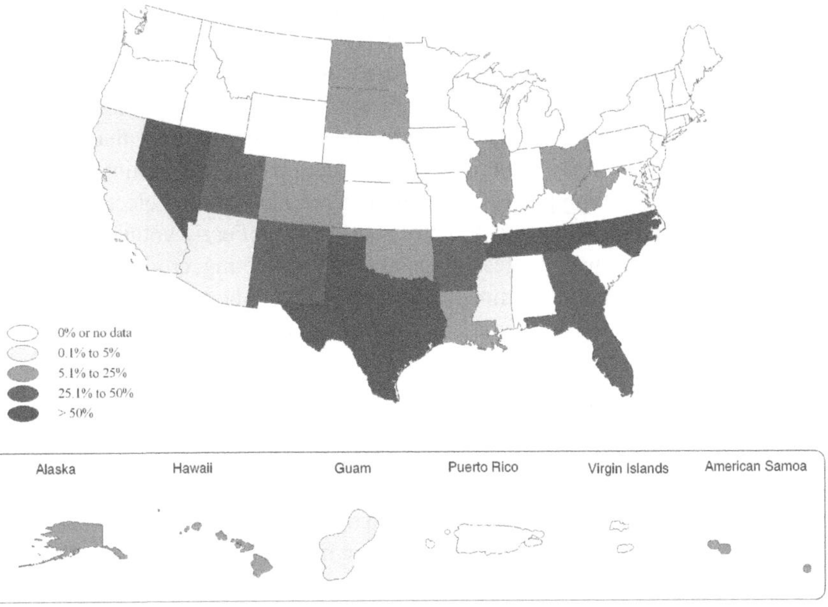

Figure 9.2 Map of Early Voting

The second component of traditional voting is through an absentee ballot, cast in a variety of means including through the mail or at an election office prior to election day. Absentee balloting dates to the Civil War, when Union states sought ways to allow soldiers to vote in the 1864 presidential election. Absentee ballot laws vary significantly across the country, with some states requiring valid excuses (such as being out of town on election day) while others allowing "no excuse" absentee balloting. In 16 of the 49 states that reported absentee voting figures, absentee voters made up 5.0 percent or less of the total voters. At the other extreme, three states (Arizona, Colorado and Washington) reported absentee voting comprised more than half of their total.

One of the more common forms of convenience voting is early voting, which allows for the casting of ballots before election day. In 2008, 24 states reported at least some level of early voting, with more than 17 million voters casting their ballots before election day. Early voting is a prominent part of the election process in Florida, Georgia, Nevada, New Mexico, North Carolina, Tennessee, and Texas; in each state, at least 30 percent of voters were reported to have cast an early ballot. Other states with lower but still significant numbers (more than 10 percent) of early voters were Alaska, Arkansas, Colorado, Hawaii, Illinois, Louisiana, North Dakota, Utah, West Virginia, and American Samoa. Early voting is most common in the southern half of the country.

Early voting changes elections in several ways. Voters who cast their ballots early forfeit the information that might have become available in the last days or weeks of a campaign. The extended time that voting is open may privilege campaigns with the resources and organizational strength to get their voters to the polls, thus making campaign financing even more important than it already is. Administratively, early voting may help election officials minimize the risks of overburdening polling places on election day in high turnout elections, and cut the time voters must wait in line to vote. Some critics of early voting contend that casting ballots in this way dilutes the social aspects of voting, eroding an important community event and thus diminishing its perceived importance. Evidence of the impact of early voting on turnout has been mixed (Gronke, Galanes-Rosenbaum and Miller, 2007).

Another convenience voting mechanism is vote-by-mail, in which entire precincts, jurisdictions or even states do away with traditional balloting entirely, and replace it with votes cast through the postal system. Voters are mailed a ballot, fill it out and return it through the mail. Oregon switched to all vote-by-mail in 1998 after the passage of a public referendum, and other states have moved at least some of their elections in that direction as well. The mode is most popular in the West: nearly all voters in Washington now vote by mail, as do many voters in California. (The EAC survey in 2008 conflated vote-by-mail with mail-in absentee voting; in 2010, the survey will separate the two options. Generally, vote-by-mail denotes the replacement of in-person voting, while mail-in absentee voting augments it).

Like early voting, vote-by-mail has its advocates and its critics. Advocates argue that it increases the convenience of voting and boosts turnout, while saving money by reducing the cost of administering elections. Vote-by-mail is also said to increase uniformity in the voting process, and the iterative process of mailing out ballots and receiving notices of non-delivery helps keep voter registration rolls up-to-date (Bradbury, 2005). Critics counter that vote-by-mail leaves voters at the mercy of a sometimes unreliable US Postal Service; does little if anything to boost turnout; is vulnerable to fraud; and privileges the affluent over low income voters (Slater and James, 2007). (On turnout and vote-by-mail, see Southwell and Burchett, 2000, and Berinsky, Burns and Traugott, 2001). While vote-by-mail has increased rapidly since Oregon adopted it 12 years ago, it remains to be seen whether its growth will continue and reach beyond the West Coast.

Same Day/Election Day Registration

While early voting and other forms of convenience voting are aimed principally at making ballot casting easier, same day registration (SDR) or election day registration (EDR) aims at lowering another barrier: voter registration. In SDR and EDR, voters are allowed to register and vote on the same day, either before or on election day itself. In addition to states that open SDR and EDR to all voters, some states offer it only to certain kinds voters, such as recently discharged uniformed

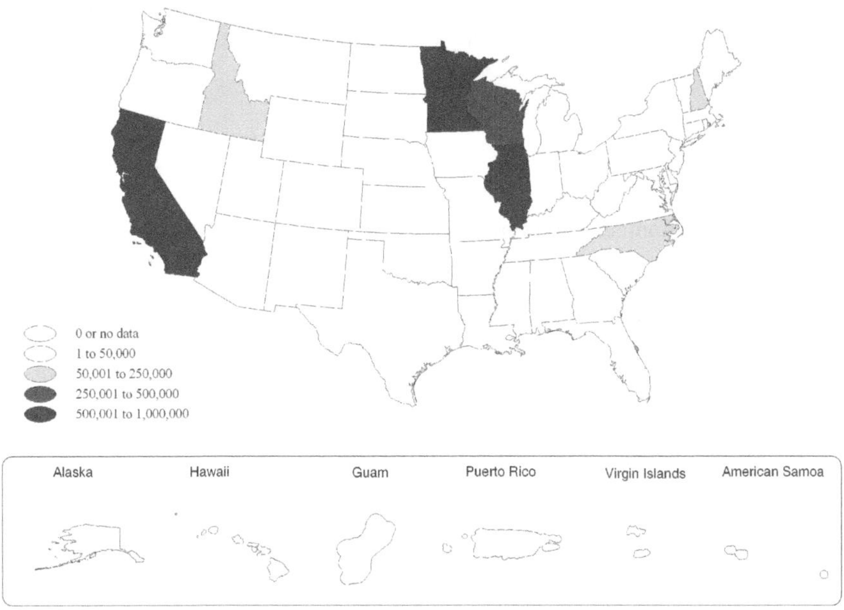

Figure 9.3 Map of Same Day/Election Day Voting Rates

services personnel or new citizens. In still others, SDR can occur as a side effect of the election calendar; for example, when the close of registration comes after the start of early voting, creating a window in which people can do both on the same day.

The 2008 EAVS asked about SDR/EDR in a roundabout way, due to refusal of some states to classify their voting as SDR or EDR, and problems with states tracking the votes of those who registered at the same time they cast a ballot. The 2008 EAVS asked states to report the number of registration forms received on days that voters could register and vote on the same day. These figures give some indication of the prevalence of SDR/EDR.

Nationally, 17 states reported receiving 3,616,874 voter registration applications on days when the voter could then cast a ballot. At least 963,144 of these applications resulted in new voters being added to the voter list; however, because most states did not categorize the applications they received, the actual number was almost certainly much higher.

Like other non-traditional registration and voting practices, SDR/EDR has been criticized as creating a vulnerability to election fraud. Without effective registration lists that can be checked against databases in an effort to detect dead or ineligible voters, critics argue, non-eligible voters can walk in, register and vote without the necessary screening mechanism of the traditional registration process. In response to these concerns, and to facilitate other innovations such as vote centers (see below), some states have adopted electronic pollbooks that update continuously during a voting period, enabling election officials to track who has voted in real-time.

Provisional Voting

Federal law guarantees voters in federal elections the right to cast a provisional ballot if they are otherwise refused a regular one. Generally, the need to vote provisionally occurs when a voter's name is not on the registration list, and the voter contends that he or she is lawfully registered to vote. Provisional ballots are used by states for a wide variety of other purposes, including the processing of change-of-address (in Ohio) and the failure by a voter to produce required ID. Another source of provisional ballots is voters who request absentee ballots, then either do not receive them or fail to return them.

In close elections, provisional ballots can be the focus of intense scrutiny, as their validity may not be determined for days or weeks after election day. In addition, some have looked to provisional ballots as an overall indicator of how well an election system works. Gerken has called provisional ballots the "miner's canary" for election administration, providing a warning sign that a particular election system is "under stress" (Gerken, 2009).

More than 2 million provisional ballots were cast during the 2008 Presidential election. Nearly three-quarters of these ballots were cast in just four states: California, New York, Ohio and Arizona (Table 9.1). Nationally, 28.2 percent of provisional ballots, or 609,016 ballots, were rejected, while another 5.5 percent were counted only in part. The most common reason for rejection, accounting for slightly more than half the rejections, was the determination that a voter was not properly registered in the state; another 16.8 percent were rejected for not being properly registered in the local jurisdiction or in the wrong precinct. Failure to provide sufficient identification resulted in the rejection of 12,321 ballots.

In spite of Gerken's assertion, interpreting provisional ballot statistics remains problematic. States employ provisional ballots in very different ways, and acceptance rates are difficult to interpret. High rates of provisional ballot acceptance may mean that voters are being improperly denied a regular ballot; low rates of acceptance may mean that many voters who think they are properly

Table 9.1 Top Provisional Ballot States, 2008

State	Provisional Ballots Cast	Provisional Ballots Rejected	% Rejected
California	798,332	136,286	17.1
New York	279,319	111,843	40.0
Ohio	204,651	39,390	19.2
Arizona	151,799	44,473	29.3

registered are in fact not. Both suggest problems in the election system, so establishing a normative basis is a challenge. In any case, provisional ballots remain a time bomb in American elections, providing fertile ground for recount controversies.

Voter ID

The introduction of more stringent identification requirements for voters was one of the most hotly contested election administration issues of the past decade. A weak ID requirement was included in HAVA, but covered only first-time voters who registered by mail and did not provide an ID when they registered. The list of acceptable IDs was also long, and included photo IDs and forms of identification such as utility bills. Pushed mainly by Republicans as a weapon against voter impersonation, voter ID laws caused considerable controversy when they were passed in Georgia and Indiana in 2005, and later in Florida and Arizona. Controversy also arose when a voter ID proposal was included among the recommendations of the Commission on Federal Election Reform (the "Carter-Baker Commission") in 2005. The Supreme Court's ruling in *Crawford v. Marion County Election Board* in 2008 validated the Indiana law, although litigation and public controversy continue. To opponents, tougher ID requirements are a partisan ploy to discourage voting among the poor, ethnic and racial minorities and the elderly, all of whom opponents claim are more likely to lack the required forms of ID.[5]

Voter ID laws are generally complicated pieces of legislation, making comparison between states difficult. The number of states requiring photo ID as a prerequisite to voting, however, remains small. The EAC survey found that only two – Florida and Indiana – require photo ID of all voters (Georgia did not respond to the Statutory Overview portion of the survey; three other states nominally require photo ID but accept an affidavit in its absence). Twenty-four of the 49 responding states required some form of ID beyond oral self-identification or provision of a signature; most of these accept non-photo ID.

Voting Technology

The notorious failure of punch cards in the Florida Presidential race in 2000 launched a federal drive to replace voting equipment: some $3 billion was spent on voting technology through the EAC and HAVA, and there was wide expectation that electronic voting machines would replace the myriad punch card, optical

5 There has also been an attempt to require proof of citizenship in order to register or vote. Fewer states have moved on this issue, Georgia and Arizona being the exceptions, although with illegal immigration a leading national issue, more states are expected to follow suit.

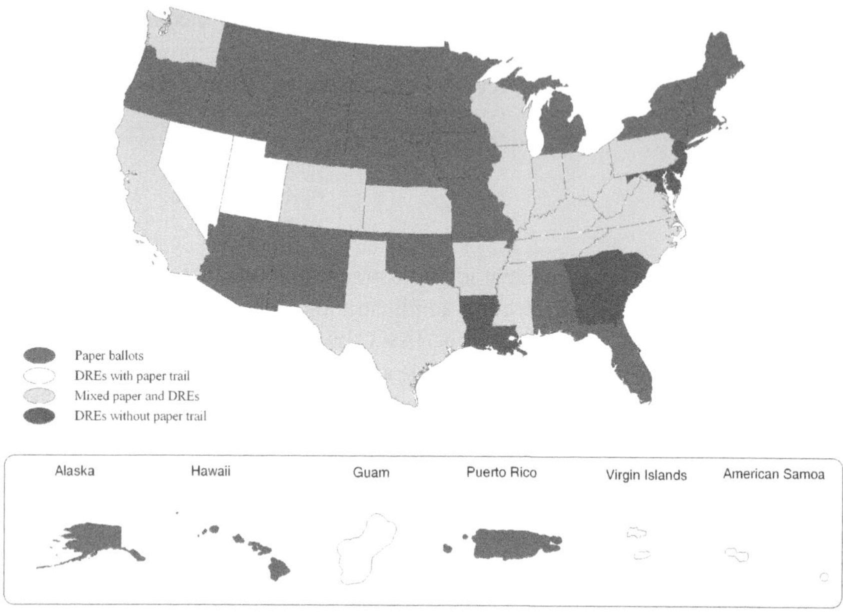

Figure 9.4 Map of Voting Technology

scan and lever machines in place across the country. Deploying direct recording electronic voting machines (DREs) turned out to be more difficult than anticipated, however. Concerns over their reliability and susceptibility to tampering spawned a cottage industry of electronic voting skeptics, who advocated for more traditional systems such as optical scan ballots that leave a paper ballot behind for recounting and auditing purposes (see Warf, 2006). Called voter-verified paper audit trails, or VVPAT, they were adopted by a series of states such, as Maryland, who had been early adopters of DREs. The 2008 election, by one estimate, was the first in which the number of voters casting ballots by DREs actually declined over the previous election (Election Data Services, 2008).

Legacy technology, local discretion and controversies such as the one surrounding electronic voting has resulted in a patchwork in the deployment of voting technology (see Figure 9.4).[6] While lever machines and punch cards were doomed by the requirements of HAVA, surviving were a wide variety of systems, including those using optically scanned ballots; various makes and models of DREs; and hybrid systems incorporating multiple technologies. Some jurisdictions

6 Because the EAC survey collects data on the number of machines rather than on how the technologies are used, I have used data from a nonpartisan advocacy group, Verified Voting, here rather than the EAC. The map in Figure 9.4 is adapted from Verified Voting's data. Available at: http://www.verifiedvoting.org/verifier/index.php?ec=standard&state=&year=.

use one technology for in-person voting, another for absentee ballots, and a third for voting by disabled persons.

States that rely on DREs without a paper record are clustered along the East Coast. With New York's decision to adopt paper ballots, lever machines have vanished from the country's electoral landscape, and punch cards are in use only in some Idaho counties. Many states use combinations of technologies, either through serving different groups of voters or through variation from county to county; the state-level map in Figure 9.4 actually understates the spatial variation in voting technologies deployed.

Emerging Issues

Changes in voting technology, convenience voting and other election administration issues will continue to be foci of election scholars over the coming decade. In addition, a number of new issues appear poised for the next decade's battles over election reform. Voter identification laws are still being introduced in state legislatures in the wake of the Supreme Court's upholding of the Indiana ID law, and many of those same states will consider changes to the way they maintain their voter registration lists. In 2010 Georgia won federal clearance for a new database matching procedure which would compare new voter registration applications against drivers' license and Social Security databases, in an effort to confirm the citizenship of new voters. Arizona has a similar law. Opponents claim the matching procedure will place an unequal burden on minority, elderly and poor voters (Keefe, 2010). Other states may consider implementing such procedures on the heels of the federal clearance.

Voting by overseas citizens and uniformed services members will continue to be a focus both of research and policy, for a number of reasons. Almost alone among the major election topics, efforts to ease participation by these voters enjoys strong bipartisan support. The federal UOCAVA also gives federal lawmakers an unusual lever over what is traditionally a local and state prerogative. Finally, new technologies may be more of a boon to distant voters than to domestic ones. State experimentation in UOCAVA voting, such as allowing voter registration over the Internet and transmitting ballots electronically, will be watched closely for potential broader application. In 2009, Congress passed the Military and Overseas Voter Empowerment Act, which will prompt further changes in how these voters cast their ballots.

Conclusions

The ways that American cast their ballots shows considerable spatial variation across the United States, and while reliable historical data is lacking, it appears that this divergence is increasing. Strong regional patterns appear as well. In the

West, voters increasingly cast their ballots through the mail. In the upper Midwest, and in selected other states, governments are lowering barriers to registration by allowing voters to register and cast their ballots on the same day. Early voting has become popular across the South and Southwest. Finally, some Southern and Western states have passed new laws that place additional burdens on voters to prove their identity and citizenship.

This divergence is important in several ways and at several scales. Nationally, the variation introduces a potentially significant variable in the ways that federal elections, including presidential ones, are decided. Combined, factors such as convenience voting, voter ID and registration rules, provisional ballots and voting technology can help mold the electorate and the number of counted ballots, and therefore can impact candidates' vote totals. Lengthening the window in which ballots are cast through early voting and extended deadlines for some voters means that people now vote with different sets of political knowledge. Late-breaking events, such as the disclosure of President George Bush's 1976 drunk driving charge, may come before election day but after substantial numbers of voters have already cast their ballots in some states. In addition, campaigns run under early voting laws may be more expensive or place a premium on a good campaign organization, which is needed to leverage the increased time to get a candidate's backers to the polls. The impact of all of these variations is still poorly understood.

At the state level, election laws are vulnerable to manipulation for partisan advantage. Laws that restrict or expand access to the ballot through raising or lowering the bar for registration and voting will rarely fall equally on all segments of the electorate. It is an article of faith among both parties that the marginal voter – the voter who will vote only in high turnout elections and even then may need additional motivation through get-out-the-vote techniques – will tend to vote Democratic in most places. Generally, Republicans, citing fear of electoral fraud, favor putting additional burdens on voters to ensure that only eligible voters cast ballots. Democrats, emphasizing the importance of broad participation and leery of laws that might disenfranchise, argue for making registering and voting easier. Both parties are well aware of the partisan impacts of changes in election laws.

Locally, changes in election laws inevitably raise questions of equal protection, particularly if discretion is left to local election officials in implementation and enforcement. Equal access to the franchise for voters of ethnic or racial minorities has been guaranteed relatively recently, and deep suspicions remain about the country's ability to preserve equal voting rights from discrimination and the push for partisan advantage. Barriers to voting can be both formal, through the imposition of laws that disproportionately impact certain groups, or informal, through the use of administrative policies that discriminate. Formal barriers may include restrictive registration or polling place laws, while informal techniques may be such seemingly mundane administrative decisions such as the distribution of voting machines or the location of polling places or early vote centers. New types of data, such as information on waiting times at polling places for white and minority voters, may be needed to measure disparate impact of some of the voting changes outlined in this chapter. In fact, given the growing divergence in American voting laws, such issues may be the new battleground for voting rights as the twenty-first century progresses.

Beyond the impact of election laws and practices on the political process, scholars interested in American regionalism can glean insights by the ways states and regions run their elections, and the discourses of reform that are produced. Debates over voter ID, for example, can help us understand the ways that meanings of citizenship are contested in different parts of the United States. The gulf between those who see voting as a right and those who see it primarily as an obligation or a privilege intersects with national and regional debates over race and illegal immigration, among other topics.

This chapter has relied upon new and improving datasets, but the push for improved data collection in still in its early stages. There is still a crying need for more and better quality data, at lower scales (including, most importantly, the precinct level), and for better and more widely available maps. Data the county (or equivalent) and state levels don't tell us much about the way variation in balloting laws and practices may discriminate against particular communities. Since few states maintain voter registration data by race, reference to demographic variables will generally be accomplished through Census data at the tract- and lower levels. Calculating these variables will require more accurate and up-to-date precinct maps, made available through election websites. Few jurisdictions provide such maps at present.

Geographers, particularly political geographers, should be attuned to the potential research and service opportunities available in helping improve election data, cartography and analysis. Research into the geography of election administration can be a valuable and engaging teaching tool, and an outlet for student and professor community outreach. Every university, community college and high school resides in an election jurisdiction, and many of these jurisdictions are eager for research and service partnerships. On the scholarly side, disputes over voting rules will demand the expertise of scholars who can evaluate their potential impact; litigation over voting rights and other election topics shows no signs of abating. Geographers are well positioned to make valuable contributions to this field.

The tug-of-war that is embedded in the current landscape of American voting and described briefly in this chapter is largely the result of two distinctly American facets of democracy: local control of election administration, and more fundamentally, longstanding differences in how Americans view democracy and the right to vote. The extreme federalism of the US election system leaves states and their local jurisdictions with wide discretion on how to run the voting process; local preferences and histories are producing different solutions to the challenges of casting and counting ballots in twenty-first century. These local practices combine with deep disagreements over the value of the right to vote and the obligations that can be placed upon a voter by the state. These differences predate the US Constitution and have never been entirely resolved: it is no accident that since the Bill of Rights, nearly half of the Constitution's 17 amendments have concerned the right to vote. Despite these amendments and the passage of such landmark federal laws as the Voting Rights Act, Americans have yet to arrive at a national consensus on who should be allowed to vote and the role of the state in making voting easier or harder to do. The divergence across the country in how Americans cast their ballots will be well worth studying in coming years.

References

Agnew, J. 1996. "Mapping Politics: How Context Counts in Electoral Geography." *Political Geography* 15, 2: 129–46.

Berinsky, A., Burns, N., and Traugott, M. 2001. "Who Votes By Mail? A Dynamic Model of the Individual-Level Consequences of Voting-By-Mail Systems," *Public Opinion Quarterly*.

Bradbury, B. 2005. "Vote-by-mail: The Real Winner is Democracy," *Washington Post*, January 1, A23.

Election Data Services. 2008. "Nation Sees Drop in Use of Electronic Voting Equipment for 2008 Election – A First." Available at: http://www.electiondataservices.com/images/File/NR_VoteEquip_Nov-2008w Appendix2.pdf.

Fallows, J. 2010. "Fallows on the News: NY Mosque and More." Commentary on All Things Considered, National Public Radio, August 14. Transcript available at: http://www.npr.org/templates/story/story.php?storyId=129201849.

Gerken, H. 2009a. *The Democracy Index*. Princeton, Princeton University Press.

Gerken, H. 2009b. "Provisional Ballots: The Miner's Canary for Election Administration." In *Provisional Ballots: An Imperfect Solution*, report published by the Pew Center on the States. Available at: http://www.pewcenteronthestates.org/initiatives_detail.aspx?initiativeID=54789.

Gronke, P., E. Galanes-Rosenbaum and P. Miller. 2007. "Early Voting and Turnout." *Political Science and Politics* 40: 639–45.

Gronke, P., E. Galanes-Rosenbaum, P.A. Miller and D. Toffey. 2008. "Convenience Voting." *Annual Review of Political Science* 11: 437–55.

Keefe, B. 2010. "Ga. policy halts voting by illegal immigrants, backers say." *Atlanta Journal-Constitution*, August 30.

King, G. 1996. "Why Context Should Not Count." *Political Geography* 15, 2: 159–64.

Overton, S. 2006. "Voter Identification." *Michigan Law Review* 105, 631.

Slater, M. and T. James. 2007. "Vote-by-mail Doesn't Deliver." Commentary on Tompaine.com. Available at: http://www.tompaine.com/articles/2007/06/29/votebymail_doesnt_deliver.php [last accessed: February 18, 2010].

Southwell, P., and J.I. Burchett. 2000. "The Effect of All-Mail Elections on Voter Turnout." *American Politics Quarterly* 28(1): 72–9.

Taylor, P., and R.J. Johnston. 1979. *The Geography of Elections.* Penguin, Harmondsworth.

Tokaji, D. 2005. "The Moneyball Approach to Election Reform." October 18 commentary. Available at: http://moritzlaw.osu.edu/electionlaw/comments/2005/051018.php.

Warf, B. 2006. "Voting Technologies and Residual Ballots in the 2000 and 2004 Presidential Elections." *Political Geography* 27: 664–77.

Chapter 10

The 2008 US Presidential Primary Elections in Geographic Perspective

Fred M. Shelley and Heather Hollen[1]

The 2008 US presidential election was one of the most riveting and fascinating presidential elections in the country's history. Two months before the general election, it was all but certain that the voters of the United States would either elect an African-American president or a female Vice President for the first time in US history.

The selections of Senators Barack Obama and John McCain as the presidential nominees of the Democratic and Republican parties, respectively, followed many months of intensive campaigning among potential nominees in both parties. Voters throughout the country expressed their preferences among these potential nominees in primary elections and caucuses in all 50 states. The results of these primary elections themselves reveal a fascinating portrait of geographic differences within states as voters expressed their preferences for the Presidential nominees of the two major parties.

The purpose of this chapter is to examine the geography of the 2008 primary elections. In doing so, it is possible to describe the history and geography of the United States' unique system of selecting major-party Presidential candidates. Detailed discussion of the 2008 campaigns for the Democratic and Republican nominations are presented along with the results of cartographic and statistical analysis of county-level voting outcomes in some key states. Furthermore, analytical issues associated with using electoral-geographic analysis of US presidential primary elections are called to attention. A general discussion on the impacts of the primary election campaign on the general election, and on future elections concludes the chapter.

US Presidential Primary Elections in Geographical and Historical Context

The process of selecting the president of the United States is a two-step process. First, both major parties select their presidential nominees. Once these nominees are chosen, they run against each other in the November general election.

1 The authors thank Trung Vinh Tran for assistance with the graphics.

Methods of selecting nominees for high public offices such as the president of the United States vary among the world's democracies. A party's nominee can be chosen in one of several ways: by party members who are elected officials, by the party elite, by party activists, and by ordinary voters (Punnett 1993). In the United States, all four methods of selecting party nominees to run for President have been used at various times over the course of American history. Over the past two centuries, major American political parties have moved from selection of presidential nominees by members of Congress to selection by party elites to selection by representatives chosen by ordinary voters. The system of presidential primary elections used today gives ordinary voters much more influence in the selection of presidential nominees than was the case in the past, or is the case in many other countries.

Relative to political parties in many other countries, US political parties are loose coalitions of often disparate interest groups. Although each party drafts a platform that expresses the policy positions of the party on major domestic and international issues, these party platforms are not binding on the party's candidates and its members. Frequently, battles within a party for its presidential nomination reflect deeper divisions within the party over the party's priorities and its underlying political philosophy. Both parties attempt, with varying degrees of success, to unite diverse economic and cultural interests that are often diametrically opposed to one another. Tensions between these interests are often especially evident during primary elections. The 2008 election was no exception.

Prior to the New Deal of the 1930s, two key elements in the Democratic Party's uneasy coalition included Southern conservatives and Northern urban dwellers. The South had been strongly Democratic since the end of the Civil War and Reconstruction. Many Southern Democratic leaders and officeholders were nativists, supporters of prohibiting the manufacturing and sale of alcoholic beverages, supporters of state's rights, and vehement opponents of civil rights for African-Americans and other minorities (Key 1949). In the North, however, many Democrats were blue-collar urban dwellers and labor union members. Many were European immigrants themselves, or the children or grandchildren of immigrants. Describing the Democrats of the 1930s, Barone (1990, 28) wrote that "The Democracy was a party of white southerners and northern Catholics, of Southern Baptist Prohibitionists and immigrant imbibers, of nativists and those who spoke no English, of teeming eastern cities and the wastelands of the Great Basin. Its members had little in common except that most of its members were not native-born white Protestants."

Barone went on to describe the constituencies of two leading Democrats in the House of Representatives at this time: Eugene Cox of Georgia and Adolph Sabath of Illinois. Barone wrote, "Two such different constituencies as the worn-out cotton lands of south Georgia and the mostly abandoned slum streets of Chicago's river wards could hardly be imagined; little wonder that their representatives, members of the same party who were yoked together on the Rules Committee, could agree on almost nothing" (224). Franklin D. Roosevelt, however, was able to create a Democratic majority during the 1930s and 1940s by linking these two traditional Democratic constituencies with African-Americans, ideological liberals, and academics and other highly educated professionals. This "New Deal" coalition elected Roosevelt four times. Even after it broke down at the presidential level, the Democrats controlled the House of Representatives for all but four years between 1931 and 1994. Although the Democrats have held the White House for only 12 of the previous 40 years, this coalition continues to impact Democratic Party politics. Tension between old-line, working class Democrats and ideological liberals and professionals has characterized the party's politics for decades, and was critical to the 2008 primary elections (Shelley 2008).

The Republicans have often been as divided as the Democrats over the past several decades. Prior to the 1960s, the Republicans were split between what some historians have termed "Wall Street" and "Main Street" Republicans. The "Wall Street" Republicans were based in the large metropolitan areas of the East and Midwest. They were internationalist in outlook, oriented to big business, and supportive of an assertive foreign policy; after World War II, they strongly supported an aggressive stance against the Soviets in the Cold War. "Main Street" Republicans, on the other hand, tended to come from small towns and rural areas west of the Appalachians. They were much more skeptical of foreign involvement and supported agrarian interests.

During the 1960s and 1970s, the Republicans drifted to the conservative side of the political spectrum. The nomination of Barry Goldwater, whose candidacy was strongly opposed by Northeastern liberal Republicans, for President in 1964 signaled the rise of conservatives, who often opposed the continued expansion of government, within the Republican Party. As conservatives became increasingly prominent within the Republican Party, they became divided into several distinct factions. Economic conservatives favored restricting government influence on the economy. Social conservatives favored increased government control over such issues as abortion, gay marriage, and immigration. Neoconservatives, who support an aggressive foreign policy stance, represent a third major interest group in the Republican Party.

Just as Roosevelt succeeded in uniting the Democratic factions in the 1930s, Ronald Reagan was successful in uniting Republicans, both economic conservatives and social conservatives, in the 1980s. Since Reagan's time, however, the Republicans have faced continuing tension between economic, social, and foreign policy conservatives. These tensions were evident in the early primaries in 2008 as well.

The task of resolving intraparty tensions and attempting to unite the disparate interests that comprise major political parties has been undertaken in different ways over the course of US political history. During the early nineteenth century, members of Congress selected their parties' nominees for President and Vice President. In 1831, however, the Anti-Masonic Party held the first national party convention to select the party's 1832 nominees. The Democratic Party soon followed suit, and the Whig Party held its first national convention in 1839. Since that time, US political parties have chosen their nominees at national party conventions.

The transition from selecting nominees by members of Congress to selecting them at national party conventions was facilitated by the development of the railroad, which enabled delegates from most parts of the country to attend them after at most a few days of travel.

Nineteenth-century nominating conventions were held at locations easily accessible by rail. The 1831 Anti-Masonic convention was held in Baltimore, as were the first six Democratic national conventions between 1832 and 1852 and five of the six Whig national conventions between 1839 and 1860. After the Civil War, Chicago emerged as the center of the nation's railroad network. Between 1860 and 1916, nine of the Republican Party's 15 nominating conventions took place in Chicago. Today, delegates from throughout the country can reach the convention site within a day. Since transportation time and cost are no longer critical factors in selecting convention sites, political or symbolic factors have taken precedence in site selection. For example, the Republicans held their 2004 convention in New York in part to remind voters of Republican President George W. Bush's response to the September 11, 2001 terrorist attack on New York's World Trade Center.

During the nineteenth and early twentieth centuries, Republican and Democratic Party conventions were dominated by party elites, including political bosses whose decisions were made in "smoke-filled rooms." The public had little input in the selection of convention delegates or party nominees. In 1910, however, Oregon established the first presidential preference primary. Oregon's delegates to the parties' national conventions were required to support the winner of the state's primary. Between 1920 and 1968, 13 to 20 states held primary elections each presidential election year (Ware 2002). However, many of these primaries were "beauty contests." In many states, delegates were selected independently of the presidential primaries, and they were under no obligation to support the candidate who got the most votes in their state's primary election.

In 1968, the Democratic Party held a bitterly contested nomination battle. In March of that year, President Lyndon Johnson almost lost the first-in-the-nation New Hampshire primary to his challenger, Senator Eugene McCarthy, who opposed continuing US involvement in the Vietnam War (White 1969; Shelley 2008). Johnson withdrew from the race soon afterwards and Senator Robert Kennedy and Vice President Hubert Humphrey entered the race for the Democratic nomination. Kennedy, who also opposed continuing US involvement in the Vietnam War, ran head-to-head in several primaries against McCarthy before Kennedy's tragic assassination after the California primary in June. Most of the delegates selected in primary elections were pledged to either Kennedy or McCarthy. Humphrey did not enter the primaries. Nevertheless, he was supported by enough delegates and party leaders from non-primary states to win the nomination. He lost the subsequent general election to the Republican nominee, Richard Nixon.

After the bitter intraparty battle of 1968, the Democrats initiated party reforms that gave greater input to ordinary voters in the selection of their party's presidential nominees. These reforms were used for the first time in the 1972 election. The Republicans initiated similar reforms in the 1970s. These new rules mandated more public participation in the selection of convention delegates. All states now hold either primary elections or caucuses in order to select convention delegates.

Primary election procedures vary between parties and from state to state. Some states hold "closed" primaries in which a voter is required to be a registered party member in advance in order to participate in that party's primary. Others use "open" primaries, in which a voter can choose either party's primary ballot at the polls and is not required to be a registered party member. Delegates chosen through primary elections or precinct caucuses are pledged to particular candidates and are required to vote for their candidate in the first round of balloting at the national convention. The methods by which the parties allocate pledged delegates between candidates for the nomination also vary between parties. The Democrats use a system of proportional representation. The number of delegates pledged to each candidate in a state's primary is proportional to that candidate's percentage of vote in the primary. In contrast, in many states the Republicans use a "winner-take-all" system in which all the elected delegates are pledged to the candidate who wins the most votes in that state's primary. About 20 percent of Democratic delegates are so-called "superdelegates." Superdelegates, who include Democratic members of Congress, state governors, and leading party professionals, are free to vote for whomever they wish and are not bound by primary results.

In 2008, the winner-take-all system worked to McCain's advantage in his quest for the Republican nomination, because McCain won several primaries over his leading opponents by narrow margins. For example, in the Missouri Republican primary on February 5 McCain got 194,053 votes (33.0 percent), Mike Huckabee got 185,642 (31.5 percent), and Mitt Romney got 172,329 (29.3 percent) (Figure 10.1). However, in accordance with the winner-take-all system McCain got all 58 of Missouri's pledged delegates. In Oklahoma, McCain got 122,722 votes (36.6 percent) as compared to 111,899 for Huckabee (33.4 percent) and 83,030 for Romney (24.8 percent) (Figure 10.2). Under Oklahoma's rules, three of the state's 25 delegates go to the winner in each of the state's five House of Representatives districts, with the other ten going to the statewide winner. Huckabee won pluralities in two of the five districts and gained six delegates, while McCain got 19 delegates by winning pluralities in the other three districts and a statewide plurality. Romney, despite finishing a close third in both Missouri and Oklahoma, won no delegates from either of these states.

Furthermore, when compared to the Democratic party, the proportional system helped Obama, who lost most of the larger states to Clinton but won a majority of primaries and caucuses in the smaller states. In Missouri, Obama had 406,917 votes (49.2 percent) with 395,185 (47.9 percent) for Clinton. Each got 36 of the state's 72 delegates. In Oklahoma, Clinton won a substantial majority, with 228,480 votes (54.8 percent) to 130,130 (31.2 percent) for Obama and 42,725 (10.2 percent)

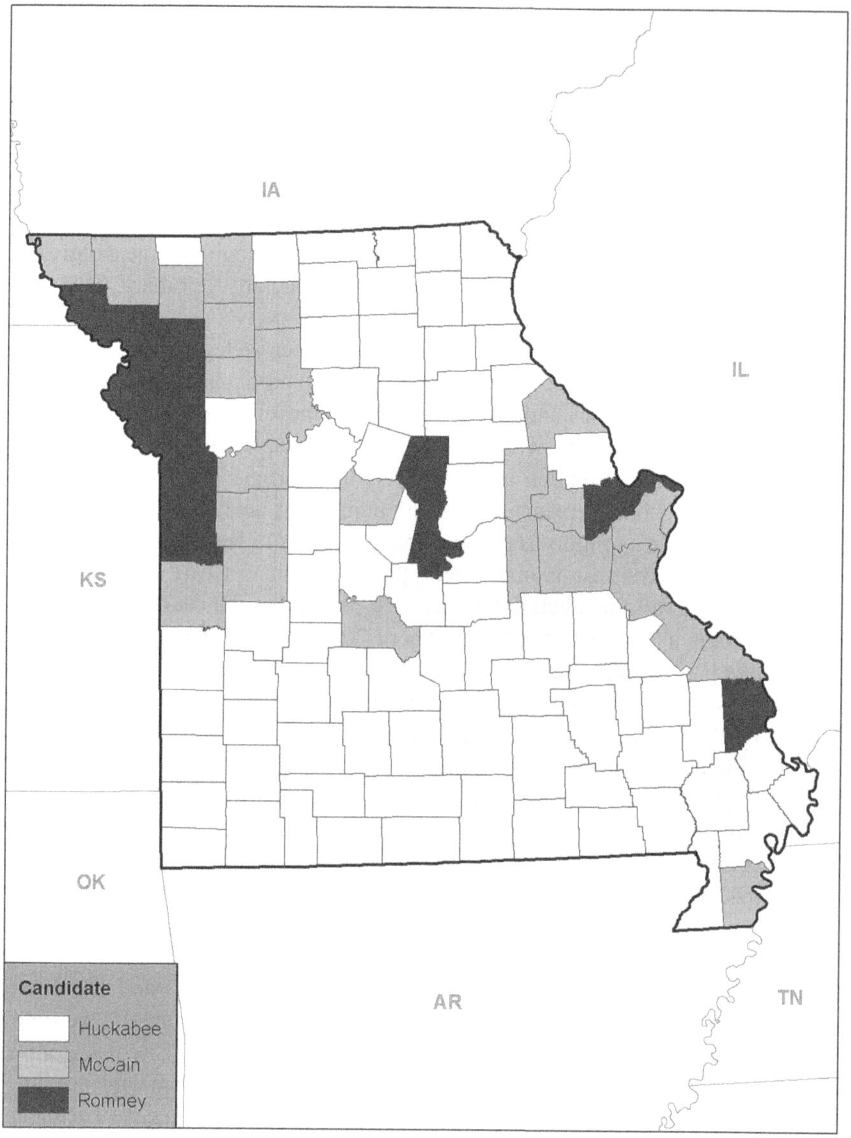

Figure 10.1 Map Showing the Results of Missouri Republican Primary, 2008

for the third-place finisher, John Edwards. Clinton won 24 of the 38 Oklahoma delegates, with 14 going to Obama. Because Oklahoma Democratic Party policy requires a candidate to achieve at least 15 percent of the vote to get any delegates, Edwards got no pledged delegates. Under Republican rules, Obama would have won all of Missouri's delegates and Clinton would have won all of Oklahoma's

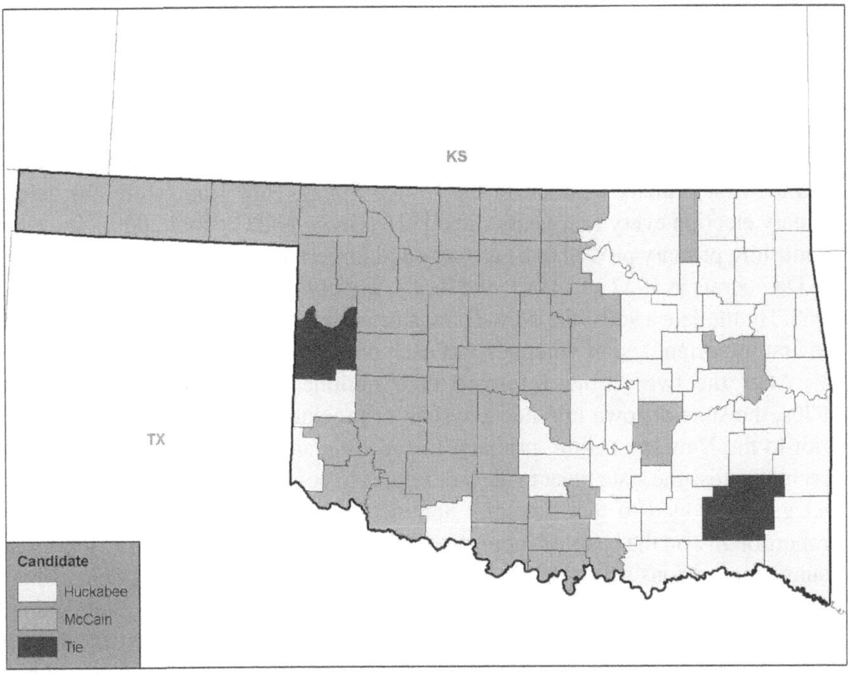

Figure 10.2 Map Showing the Results of Oklahoma Republican Primary, 2008

delegates. However, if all states had used winner-take-all systems, Obama would have lost the Democratic Party nomination to Clinton, who won many large-state primary elections including those in California, Texas, and New York.

The Primary Calendar and its Impact

The United States' primary election system is also unique in that primaries are not held simultaneously. Rather, various states hold their primary elections at different times over a period of several months. In 2008, the first primary elections took place in January; the final ones occurred in June.

The order in which the states hold primary elections is critical to the eventual outcome of the process. Candidates who do poorly in early primaries often drop out of the race. Thus, voters in states that hold later primaries may be unable to vote for their initial candidates of choice. In January 2008, for example, voters in New Hampshire, which has a monopoly over the first primary each election year, could choose among seven Republicans and six Democrats. By March, all of the viable candidates except for McCain, Obama, and Clinton had dropped out of the race. McCain had clinched the Republican nomination by early March, and Republican voters in late-voting states had no input into the selection of their party's nominee. As a general rule, the number of viable alternatives available to voters declines from earlier to later primaries (Shugart 2004).

Because states holding early primaries tend to have more influence on the outcome, there was considerable controversy about what states would be permitted to hold the earliest primaries in the 2008 cycle. Since 1952, New Hampshire has held the first primary election in the United States. New Hampshire has held a primary election every four years since 1916 (Gregg and Gardner 2000). The New Hampshire primary proved to be an early and important gauge of voter support for the Democrats in 1952 and 1968, and for the Republicans in 1952, 1960, 1964, and 1968. By the late 1960s, the New Hampshire primary had become recognized as the first important test of voter support each presidential election year.

After the two parties reformed their nominee selection processes in the 1970s, the state of Iowa initiated a system of precinct caucuses, which were held prior to the New Hampshire primary. Democratic and Republican voters in each precinct across the state meet to elect delegates to county conventions (Hull 2007). Delegates pledged to that potential nominee at the precinct level are allocated proportionally on the basis of support levels within the precinct (Squire 1989). The county conventions select delegates to Congressional district conventions, which in turn select delegates to statewide conventions. The Iowa caucus system gained nationwide attention in 1976, when then little-known Jimmy Carter, the former governor of Georgia, won a plurality. Carter's success in Iowa helped propel him to victory in the New Hampshire primary and eventually to the Democratic Party nomination and the Presidency.

Since 1976, Iowa and New Hampshire have held the first caucuses and first primary, respectively, in the country each presidential election year. In 1977, the New Hampshire legislature passed a law requiring New Hampshire's primary to precede those of any other state (Gregg and Gardner 2000). The parties were authorized to move the New Hampshire primary to an earlier date if another state announced plans to hold its primary earlier than that of New Hampshire. On several occasions, New Hampshire has moved its primary earlier and earlier as other states have scheduled early primaries also. In 1968, the New Hampshire primary was held on March 12; in 2008, it took place on January 8. Critics have pointed out that Iowa and New Hampshire with their small populations, few minority voters, and lack of large metropolitan areas are unrepresentative of the US electorate as a whole. It is also argued that the early scheduling of Iowa's caucuses and New Hampshire primary gives these states a disproportionate influence in the outcome of presidential elections. Since 1952, only Bill Clinton in 1992 has won the Presidency without winning either the Iowa precinct caucuses or the New Hampshire primary. Moreover, Clinton did not contest the Iowa precinct caucuses that year because one of his rivals for the Democratic Party's presidential nomination that year was an Iowan, Senator Tom Harkin. Clinton and George W. Bush in 2000 were the only Presidents since 1960 to have won the general election without winning the New Hampshire primary. In 2008, Obama and Huckabee won Iowa, while Clinton and McCain won New Hampshire.

In response to the perceived disproportionate influence of Iowa and New Hampshire in the electoral process over the years, party leaders in other states considered holding their primaries earlier, or working with leaders in neighboring states to hold their primaries on the same day in order to increase the influence of their regions on the nomination process. In the 1980s, some southern Democratic leaders became concerned that Southern candidates would be uncompetitive in the Northern states of Iowa and New Hampshire. In response, they decided to hold their primaries on the same day with the idea that a Southern-oriented candidate who did well in several states at the same time would get a boost toward the Democratic Party's presidential nomination. Thus the "Super Tuesday" concept was born, and was used for the first time in 1988. In that year, the Southern primaries were divided among several candidates and a Northern candidate, Governor Michael Dukakis of Massachusetts, won the Democratic nomination. Four years later, however, Clinton swept the Super Tuesday states in 1992 en route to the Democratic nomination and victory in the general election.

Before the 2008 election cycle began, party leaders in other states also explored the idea of adjusting the primary calendar in order to give voters in their states more influence in the nomination process. In August 2006, the Democratic National Committee agreed to allow South Carolina and Nevada, along with Iowa and New Hampshire, to hold "early" primaries prior to February 5. South Carolina and Nevada are Southern and Western states, respectively, and both contain larger minority populations than do Iowa or New Hampshire. Between them, the four "early" states were seen as representative of the country as a whole in terms of both geography and demography. The Committee ruled that these four states could hold their primaries prior to February 5, but that no other state could do so.

In 2007, this ruling was challenged in Florida and Michigan. The Michigan Legislature, without the support of the Democratic National Committee, passed a bill scheduling that state's primary for January 15. Florida scheduled its primary for January 29. In response, the Committee stripped the Michigan and Florida Democratic Parties of their delegates, rendering the primaries meaningless. The Democratic candidates agreed not to campaign in these states, and Obama and several other candidates had their names taken off Michigan's Democratic ballot. Nevertheless, Democratic voters in both states participated in the primaries, both of which were won by Hillary Clinton. Later in the campaign, Clinton would argue for seating delegates from these states. As the nomination campaign drew to a close and Obama's nomination was assured, the Committee eventually agreed to seat the delegates from the two states.

As the controversy associated with the Florida and Michigan primaries continued, other states scheduled their primaries. Twenty states including California, New York, and Illinois chose February 5, the first day that the Democratic National Committee set aside for primary elections. Thus, February 5 became the 2008 version of Super Tuesday. The remaining 26 states held primaries at various times between February 9 and June 3.

The 2008 Candidates and Outcomes

As the 2008 primary season approached, potential nominees in both parties announced their intentions to run for the Presidency. The Republicans anticipated a protracted battle between several candidates, each of whom brought very different strengths. Meanwhile, Clinton was seen as the front-runner for the Democratic nomination throughout 2007. Most experts expected her to clinch the Democratic nomination early, whereas they expected a drawn-out battle for the Republican nomination. As events turned out, the Republican nomination was decided quickly whereas the Democratic nomination was not decided until the end of the primary season in June.

This uncertainty occurred in part because neither the incumbent President nor the incumbent Vice President sought the Presidency. For only the second time since 1928, both parties knew going into the 2008 presidential election campaign that neither President Bush nor Vice President Dick Cheney would be on the 2008 general election ballot. In 1951, the 22nd Amendment, which limits Presidents to two terms, was adopted. Between 1951 and 2000, only three Presidents – Dwight D. Eisenhower in 1960, Ronald Reagan in 1988, and Bill Clinton in 2000 – completed two four-year terms. Vice Presidents Richard Nixon, George H.W. Bush, and Al Gore, respectively, won their parties' presidential nominations. Bush was elected in 1988, while Nixon and Gore narrowly lost to John F. Kennedy and George W. Bush respectively. The younger Bush's Vice President, Dick Cheney, had low popularity ratings and had faced a long history of health problems. Cheney said repeatedly during Bush's Presidency that he had no interest in running for President (Barnes 2005), and he made no effort to contest the 2008 Republican nomination.

With neither George W. Bush nor Cheney on the ballot, the race for the Republican nomination was wide open. During 2007, several candidates announced plans to run. These included Senator John McCain of Arizona, former New York City Mayor Rudolph Giuliani, former Governor Mike Huckabee of Arkansas, former Governor Mitt Romney of Massachusetts, former Senator Fred Thompson of Tennessee, and Representative Ron Paul of Texas. In addition to these potential nominees, several other candidates including Senator Sam Brownback of Kansas and Representatives Duncan Hunter of California and Tom Tancredo of Colorado also initiated campaigns. None of these campaigns generated momentum, however, and all three dropped out early in the process.

Each of these candidates appealed to a very different constituency, reflecting the long-standing tension between economic, social, and foreign policy conservatives within the Republican Party. McCain, who had contested and lost the Republican nomination to Bush in 2000, emphasized his role as a maverick and an independent voice. McCain was a strong supporter of the war in Iraq, but took a more moderate position on many social and environmental issues than did many Republican voters. Giuliani, who had been Mayor of New York at the time of the 9/11 terrorist attack, emphasized national security issues. Huckabee, an ordained

Southern Baptist minister, emphasized social conservatism. Romney emphasized his business experience and the economy in his campaign, while Thompson tried to appeal to the Republican Party's conservative base. Paul, who had been the Libertarian Party's presidential nominee in 1988, argued against the war in Iraq and against government interference in the economy and in social issues.

Despite this vigorous competition within the Republican Party, many observers expected that the Democrats would win the 2008 election. Several factors appeared to boost the Democrats' chances. Since the adoption of the 22nd Amendment, a party had won three consecutive elections only once – in 1988, when George H.W. Bush succeeded Reagan. However, Reagan had won landslide victories in 1980 and 1984 and he left office four years later as a highly popular incumbent. The younger Bush, however, had been elected only narrowly in 2000 and 2004 and his popularity among voters eroded throughout his second term. By the beginning of 2008, the unpopular war in Iraq had been going on for nearly five years. Many Americans believed that the economy was weakening and expected it to continue going downhill. Corruption scandals forced the resignation or humiliation of several Republican members of Congress. In part because of these issues, Bush's popularity rating among the American electorate remained low throughout most of his second term.

Throughout most of 2007, many commentators and party professionals expected Senator Hillary Clinton to receive the Democratic nomination. However, the perceived likelihood of a Democratic victory in 2008 induced several other candidates to contest the Democratic Party's presidential nomination. The Democratic field included Senators Joseph Biden of Delaware, Christopher Dodd of Connecticut, and Barack Obama of Illinois, former Senator John Edwards of North Carolina, Governor Bill Richardson of New Mexico, and two fringe candidates, Representative Dennis Kucinich of Ohio and former Governor Mike Gravel of Alaska.

Biden, Dodd, and Richardson brought extensive governmental experience to their campaigns. Biden, who would eventually be selected as Obama's running mate, had served in the Senate for 35 years and was Chair of the Senate Foreign Relations Committee. Dodd had served in the Senate since 1980 and had previously served in the House of Representatives. Richardson had also served in the House of Representatives for 14 years, and had also served as the US representative to the United Nations and as Secretary of Energy under President Bill Clinton.

Clinton, Obama, and Edwards had had much less formal governmental experience. Although Clinton emphasized her experience during her campaign, her formal governmental experience was limited to seven years in the Senate. Edwards had held public office for only six years, serving a single term in the Senate before running unsuccessfully for Vice President in 2004. Obama had been elected to the Senate in 2004, and his previous political experience had been limited to eight years in the Illinois State Senate. Nevertheless, as 2007 ended polls showed that Clinton, Obama, and Edwards were most popular among Democratic voters. Obama and Edwards in particular emphasized the need for change in their campaigns, and it became clear as the campaign continued that the desire for change resonated more with many voters than did the experience factor. Obama also emphasized his opposition to the war in Iraq, criticizing Biden, Clinton, Dodd, and Edwards for having voted in favor of authorizing expenditures to support it in 2003. (At that time, Obama had not yet been elected to the Senate.)

The Iowa caucuses and the New Hampshire primary began to winnow down the field. Huckabee, with strong support from evangelical voters, won the most votes and delegates in the Republican caucuses in Iowa on January 3, 2008. Huckabee won 34 percent of the vote and carried 74 of Iowa's 99 counties. Romney finished second with 25 percent of the vote and carried 24 counties. On the Democratic side, Obama won with 38 percent of the vote. Edwards with 29.7 percent edged Clinton with 29.4 percent for second place. Richardson, Biden, and Dodd split less than three percent of the vote and dropped out of the race.

The New Hampshire primary took place five days later, on January 8. On the Democratic side, a poll taken shortly after Obama won the Iowa caucuses showed him with a 13-point lead over Clinton (Page 2008). However, Clinton won a narrow victory over Obama. Clinton won 39 percent of the vote, Obama won 36 percent, and Edwards won 16 percent (Shelley 2008). Later in January, Obama easily defeated Clinton and Edwards in South Carolina, where Edwards had been born. He subsequently left the race, leaving Clinton and Obama as the last contenders for the Democratic nomination. Meanwhile, McCain defeated Romney, 37 percent to 32 percent, in the Republican primary. Huckabee followed with 11 percent, and Giuliani finished fourth with 9 percent. A week later, the Republicans held a primary in Michigan. Romney, a Michigan native whose father had once served as that state's Governor, defeated McCain by a 39 percent–30 percent margin. McCain also won the primary in Florida, where Giuliani had campaigned extensively but got few votes. Shortly after the Florida primary, Giuliani dropped out of the race and endorsed McCain.

Twenty states held primaries on "Super Tuesday" on February 5. McCain won several key Republican primaries and took a commanding lead in the delegate count. After Super Tuesday, McCain's nomination was a foregone conclusion. However, Huckabee remained in the race for a few more weeks until McCain had won enough delegates to formally clinch the nomination. Attention then shifted to the Democratic Party. Clinton and Obama divided the Super Tuesday states. Clinton won primaries in California, New Jersey, Massachusetts and her home state of New York, while Obama won in Missouri, Georgia, Connecticut and his home state of Illinois. Smaller states that also voted on Super Tuesday divided evenly between Clinton and Obama, and political commentators declared Super Tuesday a draw.

After Super Tuesday, Obama picked up momentum by winning primaries in Louisiana, Maryland, Virginia, the District of Columbia, and Wisconsin throughout the month of February. Obama's success encouraged some superdelegates who had originally supported Clinton to reconsider their commitments. Clinton rebounded with victories in Ohio and Texas on March 4, while Obama won Vermont and Rhode Island on the same day. By this time, Obama had a narrow but significant lead in the Democratic delegate count. Clinton won Pennsylvania in late April and narrowly won Indiana two weeks later, but Obama won a solid victory in North Carolina on the same day. By this time, a majority of the Democratic superdelegates had declared support for Obama, increasing his lead in the Democratic delegate count. Even though Clinton won late primaries in West Virginia, Kentucky, and South Dakota, Obama was too far ahead in the delegate count for Clinton to catch up. Obama sewed up the nomination on June 3 with a victory in Montana, and Clinton formally suspended her campaign the following day.

Analysis

Numerous studies have been devoted to cartographic and statistical analysis of mapped results of individual elections or electoral sequences. Over many years, this line of research has provided valuable insights into understanding the political economies and cultures of the United States and other countries. Of course, the shortcomings of electoral geography have long been recognized, including the possibility of the ecological fallacy, associated difficulties in assuming or determining motivations of voters on the basis of electoral data, and the assumption that explanatory variables represent meaningful surrogates for the characteristics of individual voters.

These difficulties associated with interpretation of geographically disaggregated election results may be even more problematic when applied to US primary elections, for several reasons. Part of the problem lies with the nature of the primary election process itself. As we have seen, the 2008 primary season lasted five months. Issues of concern to voters in early primaries receded by later primaries, and vice versa. Those candidates who did not do well in early primaries dropped out of the race, reducing the set of choices available to voters who happened to live in states holding later primaries. The Republican race was essentially over after Super Tuesday, after which it was clear that McCain would be the Republican nominee regardless of the outcomes of later primaries. Thus, few insights can be gained from analysis of the later Republican primaries because their outcomes had no effect on the overall race for the Republican presidential nomination.

Another problem associated particularly with primary elections is that the electorate participating in each primary represents only a subset of a place's total electorate. A voter can participate in the Democratic primary or the Republican primary, but not both. The demographic, ethnic, and economic characteristics of voters participating in each party's primary can vary considerably. For example, according to the 2000 US census about 11.5 percent of Missouri's population was African-American. Because a large majority of African-Americans in Missouri and other states are Democrats, according to exit polls African-Americans represented nearly 20 percent of the Democratic Party's primary electorate in Missouri. However, less than 3 percent of the Republican voters of Missouri were African-Americans. Because the overwhelming majority of Missouri's Republican primary voters were Euro-Americans, using race as an explanatory variable to explain levels of support for Republican candidates is useless. Data from the US Census and other government sources describe the characteristics of a place's population but do not break these data down between Democratic and Republican primaries.

Recognizing these pitfalls, we nevertheless conducted county-level analysis of Republican and Democratic primary elections in various states. Statistical analyses, comparison of counties that were carried by different candidates, and state-level exit polls were used to shed light on geographic differences within and between states in primary election outcomes. Through detailed analysis of the Super Tuesday primary elections in Missouri and Oklahoma it was possible to make connections between the geographical make-up of the voters and their constituent selections. These states were chosen because both primaries were very closely contested between three leading Republican contenders – McCain, Huckabee, and Romney.

Although Missouri and Oklahoma share a boundary, Missouri has long been a bellwether. Its electoral votes have gone to the national winner of every presidential election since 1904, with the exceptions of 1956 and 2008, in which McCain won Missouri's electoral votes by a margin of less than 4,000 popular votes. Missouri's demography closely parallels that of the US as a whole. Missouri is 11.5 percent African-American. Its per capita income in 1999 was $19,936; 13.3 percent of its residents were over 65, and 13.0 percent were living in poverty in 2000. These figures for the US as a whole were $21,387, 12.4 percent, and 12.7 percent respectively. Missouri's population density of 81.2 persons per square mile is similar to the 79.6 persons per square mile across the United States. Thus, the demography and economic status of Missouri closely parallel that of the US as a whole. Oklahoma, on the other hand, is a strongly "red" state that has supported the Republican nominee for President in every election since 1952, with the exception of 1964. Relative to Missouri, Oklahoma is more rural, has fewer minorities, and has a lower per capita income.

Although McCain won both states' Republican primaries narrowly, in both cases Huckabee won a majority of counties. Huckabee won a plurality in 72 counties in Missouri, with 30 for McCain and 13 for Romney (Figure 10.1). Counties carried by all three candidates in Missouri were geographically concentrated. Nine of Romney's 13 counties were located in the Kansas City metropolitan area. McCain was strongest in more rural counties around Kansas City, and in the St. Louis metropolitan area. Huckabee swept the southern portion of Missouri (which borders his home state of Arkansas) and in northeastern Missouri, a region sometimes known as "Little Dixie" because of the Southern origins of many of the region's early settlers (Hurt 1992).

Counties carried by the three candidates in Missouri show very different characteristics. The counties carried by Romney tended to grow more rapidly and contain younger, better-educated, and higher-income voters. The median per-capita income of Romney's counties was $19,844 as compared to $16,197 for McCain and $14,885 for Huckabee. The median percentage of college graduates was 23.4 percent for Romney counties as compared to 12.1 percent for McCain counties and 10.8 percent for Huckabee counties. The median population density was 175.7 persons per square mile in Romney counties compared to 29.5 for McCain counties and 29.0 for Huckabee counties. Thus, Romney did better in urbanized as opposed to rural areas. These data suggest that Romney's appeal in Missouri was greatest among younger, more upscale, and better-educated Republicans. This is corroborated by positive correlations between county percentages for Romney with per capita income (r=0.68) and percentage of residents holding college degrees (r=0.57). Levels of support for Bush in 2004 were similar among all three groups of counties; the median Bush percentage in 2004 was 64 percent for Huckabee counties, 62 percent for McCain counties, and 59 percent for Romney counties.

In Oklahoma, Huckabee outpolled McCain in 39 of the state's 77 counties, with McCain winning the remaining 38 (Figure 10.2). Romney, who came in third, won no counties in Oklahoma. Huckabee was strongest in the poorer southern and eastern portion of the state, while McCain did best in the wealthier urban areas and in the rural western part of the state. As in Missouri, McCain's counties tended to have higher incomes and levels of education. McCain also did better among elderly Republican voters, as demonstrated in exit polls. However, the counties carried by McCain were more strongly pro-Bush in 2004, in contrast to Missouri where Huckabee's support was correlated positively with support for Bush.

On the Democratic side, the closest Super Tuesday primary was in Missouri, which Obama carried by a margin of only 11,000 out of more than 800,000 votes cast. Interestingly, Obama won his majority in Missouri despite carrying only five of the state's 114 counties along with the independent city of St. Louis. The five counties carried by Obama included St. Louis County, which includes many of the inner suburbs of St. Louis; Jackson County (Kansas City); Boone County (Columbia and the University of Missouri); Cole County (the state capital of Jefferson City), and Nodaway County (Maryville and Northwest Missouri State University). Obama's margins in the City of St. Louis, St. Louis County, and Jackson County – the three most populous local jurisdictions in the state – overcame Clinton's majority elsewhere.

The Obama counties contain substantial populations of African-Americans, students, and professionals who made up the core of Obama's support. The City of St Louis, which Obama won by a 71–29 margin, is 50.5 percent African-American, and St Louis and Jackson Counties are both over 20 percent African-American. St Louis County is Missouri's highest-income county, with a per capita income of $27,595. Statewide, the county-level correlation between the

Obama vote and per capita income was 0.63, and the correlation with population density was 0.54. In Missouri, Clinton did best in lower-income counties with less educated populations – a pattern that would become even more clear in later primaries.

As we have seen, Obama surged ahead in the weeks after Super Tuesday, winning primaries in several states later in February. In our analysis, we focused especially on the states of Virginia and Wisconsin. Obama won the Virginia primaries, along with primaries in neighboring Maryland and the District of Columbia, on February 12. He won the Wisconsin primary a week later, on February 19. In Virginia, Obama got 627,820 votes (64 percent) as compared with 349,766 for Clinton (35 percent) (Figure 10.3). Obama won Wisconsin over Clinton by a vote of 646,851 (58 percent) to 453,954 (41 percent) (Figure 10.4).

Although Obama won both Virginia and Wisconsin decisively, the geographic patterns of support for Obama and Clinton in these two states were strikingly different. Virginia's Democratic electorate was highly polarized. Obama won the primary by sweeping the state's most populous areas including the suburbs of Washington, DC, and the Richmond and Norfolk/Virginia Beach metropolitan areas. Clinton won large majorities in rural areas of western Virginia. Statewide, Virginia has 114 counties and independent cities. Obama won 75 percent or more

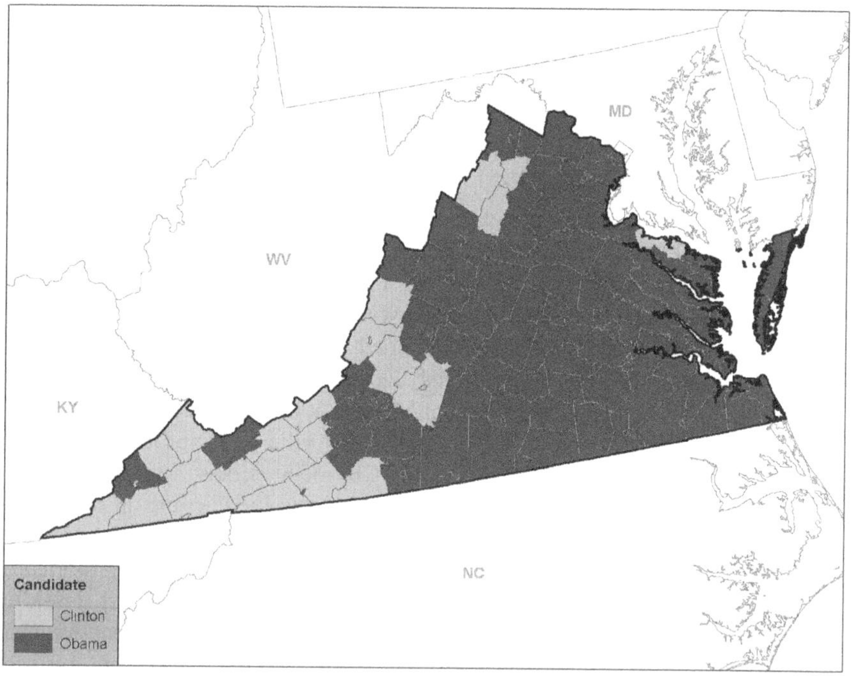

Figure 10.3 Map Showing the Results of Virginia Democratic Primary, 2008

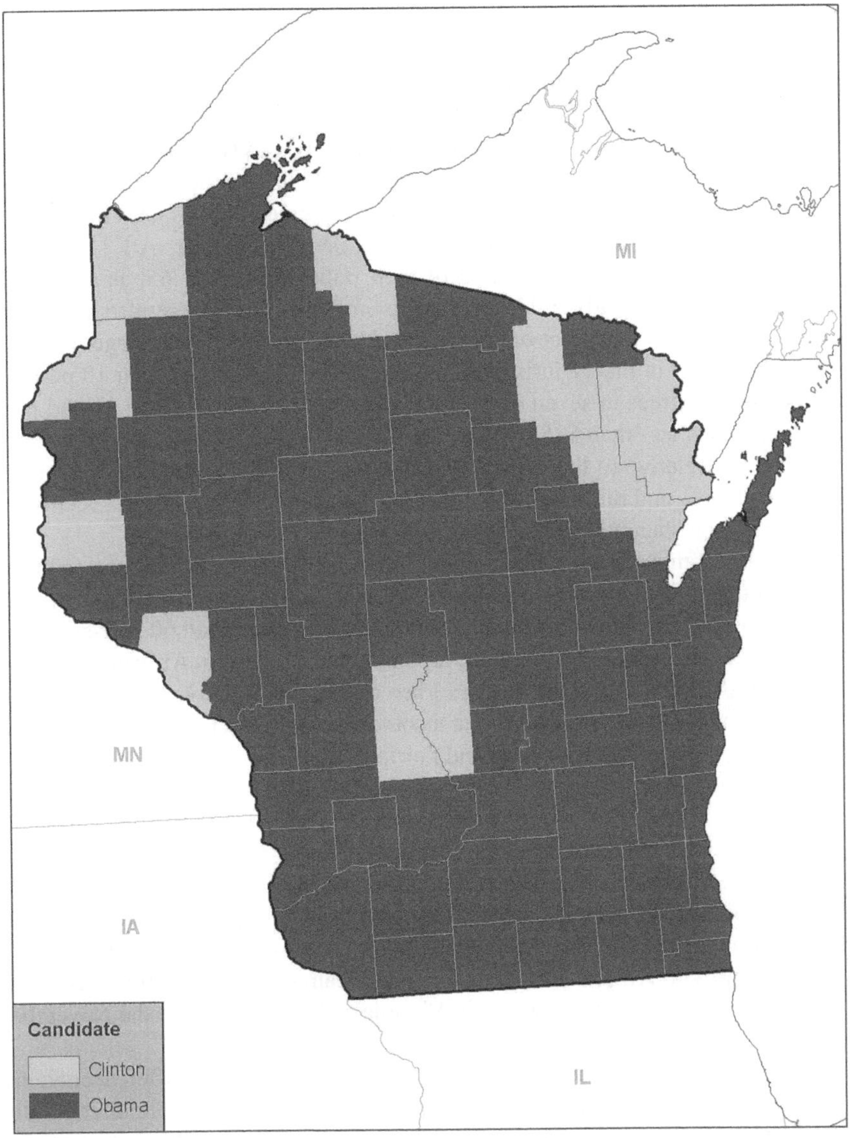

Figure 10.4 Map Showing the Results of Wisconsin Democratic Primary, 2008

of the vote in 14 of these jurisdictions, and he won less than 25 percent of the vote in 11 others. Most of the 14 Obama jurisdictions have large African-American populations. Virginia's overall population was 19.9 percent African-American in 2000. All but one of the 14 Obama jurisdictions were at least 43 percent African-American. The exception was the city of Charlottesville, home of the University of

Virginia. All but one had fewer elderly persons than the statewide average. Eight of the 14 were central cities with high population densities (Charlottesville, Danville, Emporia, Hampton, Newport News, Petersburg, Portsmouth, and Richmond). Only two of the 14 were carried by Bush in 2004.

The 11 Clinton jurisdictions included nine counties and two small independent cities, Norton and Wise. All of these are located in the Appalachian area of western Virginia. The population densities of the nine counties are all less than half the statewide density of 178.8 persons per square mile. All 11 Clinton jurisdictions had at least 14 percent of their populations over 65, as compared to 11.6 percent statewide. Education levels in these places were also strikingly low. Statewide, 29.5 percent of Virginia residents over 25 held college degrees; the highest total in the 11 jurisdictions was 14 percent and less than 10 percent held college degrees in seven of the remaining ten jurisdictions. All 11 had per-capita incomes less than $18,000, as compared to a statewide per capita income of $23,975. Relative to the rest of Virginia, the Clinton jurisdictions were low-income, elderly, and rural with low education levels and, in most cases, economic and population stagnation.

We also compared the Obama and Clinton jurisdictions to the four counties (Arlington, Fairfax, Loudoun, and Prince William) that include most of the suburbs of Washington, DC. Obama won all four easily, with more than 60 percent of the vote in each despite the fact that none are more than 10 percent African-American except for Prince William, the most exurban of the four. Relative to the rest of the state, these places have high per capita incomes and high levels of education. More than half of the adults in Arlington and Fairfax Counties held college degrees, and per capita income in these counties was more than $37,000 per year. Loudoun County, which has grown by 58 percent since 2000, is one of the fastest-growing counties in the United States. Not surprisingly, only 5.6 percent of Loudoun's residents are over 65. These data, along with data from the strongest Obama and Clinton jurisdictions, confirmed that Obama's base of support in Virginia came from upscale, well-educated voters along with African-Americans. These upscale voters along with African-Americans helped Obama become the first Democrat since Lyndon Johnson in 1964 to win Virginia's electoral votes in the November general election.

The Wisconsin primary, held a week later, yielded a strikingly different outcome when examined geographically. In contrast to Virginia, there was relatively little difference between counties in the percentage of Obama's vote. Obama's strongest counties were Dane County, home of the state capital of Madison and the University of Wisconsin-Madison, and Milwaukee County, whose population includes nearly two-thirds of the state's approximately 300,000 African-Americans (who make up about a quarter of the county's population). Obama won 68 percent of the vote in Dane County and 64 percent in Milwaukee County. His weakest counties were rural Douglas County (42 percent) and Forest County (43 percent). Obama carried 62 of Wisconsin's 72 counties, but won more than 60 percent of the vote in only seven of them.

Despite much smaller differences in vote percentage between counties in Wisconsin relative to Virginia, there were similar patterns of demography between the two states. All ten counties carried by Clinton had lower population densities, per capita incomes, and percentages of college graduates, and higher percentages of elderly residents, than statewide totals. The five counties carried by Obama with more than 60 percent of the vote, in addition to Dane and Milwaukee, included counties containing three other branches of the University of Wisconsin system (Eau Claire, Oshkosh, and Whitewater). Except for the large campuses in Madison and Milwaukee, these are three of the four largest campuses in the state university system. All but one of the seven Obama counties had fewer elderly people than the statewide total and all but two had higher per capita incomes. In a much more subtle way, the Wisconsin data reflect Obama's strength among younger voters, African-Americans, and in urban and upscale communities. Clinton did better among less upscale, less educated, less wealthy, and older voters.

This pattern was even clearer in Pennsylvania, whose Democratic primary took place on April 22 after six weeks without any primary elections at all. Clinton won Pennsylvania easily, getting 1,275,039 votes (55 percent) to 1,061,441 (45 percent) for Obama. Clinton won 60 of Pennsylvania's 67 counties. The seven Obama-majority counties included Philadelphia County, suburban Delaware and Chester Counties, Centre County (State College and Pennsylvania State University), and Dauphin County (the state capital of Harrisburg). Obama's strength among young voters, African-Americans, and upscale voters was evident in these places as well. In contrast, Clinton won 72 percent or more of the vote in 12 counties, many of which were located in rural central Pennsylvania or in the Pittsburgh metropolitan area. All but one of these counties had more elderly people than the statewide total of 15.2 percent. All but one have lost population since 2000, and all 12 have less than the statewide African-American percentage of 10.7 percent. All 12 also have lower per capita incomes than the state per capita income of $20,880. Thus, Clinton's strength was greater among more rural, less educated, and older voters in Pennsylvania (York 2008). This was reinforced later in the campaign when she won the more rural states of West Virginia and Kentucky easily.

Conclusion

Primary elections provide a useful and fascinating but little explored source of information about the geography of the American electorate, although the peculiarities associated with the American primary election system can limit their usefulness. Nevertheless, state-by-state comparison of factors associated with primary election outcomes can shed valuable light on voter preferences among candidates for major-party nominations. The results of analyses of primaries in Missouri for both major parties, Oklahoma for the Republicans, and Virginia and Wisconsin for the Democrats demonstrated the appeal of Hillary Clinton and Mike Huckabee to rural, downscale voters whereas more upscale and urbane voters tended to support Barack Obama and John McCain, who eventually won their respective parties' nominations. In future elections, primary elections should continue to provide electoral geographers with valuable information about voter preferences.

References

Barnes, F. 2005. President Cheney? *The Weekly Standard*, 10:23. Available at: http://www.weeklystandard.com/Content/Public/Articles/000/000/005/295zkwpw.asp.

Barone, M. 1990. *Our Country: The Shaping of America from Roosevelt to Reagan*. New York: Free Press.

Gregg, H., and W. Gardner. 2000. *Why New Hampshire? The First-in-the-Nation Primary State*. Enfield, NH: Enfield Publishing and Distribution Company.

Hull, C.C. 2007. *Grassroots Rules: How the Iowa Caucus Helps Elect American Presidents*. Palo Alto, CA: Stanford University Press.

Hurt, R.D. 1992. *Agriculture and Slavery in Missouri's Little Dixie*. Columbia, MO: University of Missouri Press.

Key, V.O., Jr. 1949. *Southern Politics in State and Nation*. New York: Vintage Books.

Page, S. 2008. Obama, McCain Leap Ahead in New Hampshire. *USA Today*, January 6.

Punnett, R. 1993. Selecting the Party Leader in Britain: A Limited Participatory Revolution. *European Journal of Political Research*, 24, 257–76

Shelley, F.M. 2008. Sections and Elections: Reflections on the Early 2008 US Presidential Primaries. *Political Geography*, 27, 505–8

Shugart, M.S. 2004. The American Process of Selecting a President: A Comparative Perspective. *Presidential Studies Quarterly*, 34, 632–55.

Squire, P. (ed.) 1989. *The Iowa Caucuses and the 1988 Nominating Process*. Boulder, CO: Westview Press.

Ware, A. 2002. *The American Direct Primary*. New York: Cambridge University Press.

White, T.H. 1969. *The Making of the President 1968*. New York: Atheneum.

York, B. 2008. Hillary in Bitterland: Clinton Finds Her People in Pennsylvania. *National Review Online*. Available at: http://article.nationalreview.com/355050/hillary-in-bitterland/byron-york.

Chapter 11
Electoral Alignments and Place-Based Cleavages in Statewide Votes in Alabama

Nicholas Quinton and Gerald R. Webster

Geographers have fixated upon contextualized accounts of space as leaders in the broader spatial turn in the social sciences (Warf and Arias 2008). This change in approach has been enlivened by calls to link various and potentially contradictory approaches to social inquiry. In particular, geographers have shown interest in the connection between modern hegemonic approaches to social inquiry and critical (though modern) interpretations of hegemony, and what post-structural social investigation means to both. Barnes (2009) argued that there is no fundamental reason why quantitative methodologies cannot be applied to critical geography. The argument was simply that decidedly hegemonic modern scientific approaches can (and should) be applied with critical interpretations of the world in a conciliatory approach to social inquiry.

The turn to conciliatory approaches has been embraced by recent changes in the practice of electoral geography. Electoral geographers have combined traditional statistical methodologies with qualitative discussions of context in mixed-methods approaches (Brown, Knopp, and Morrill 2005; Chapman, Leib, and Webster 2007; Morrill, Knopp, and Brown 2007; Shin and Agnew 2002, 2008; Webster, Chapman, and Leib 2010; Webster and Leib 2001, 2002). The general application of a conciliatory mode of social inquiry is central to the present work. In particular the connection between quantitative approaches and context centered investigation is addressed in the linkage between particular elections and their spatial vote patterns.

Accounting for Context

During the past two decades electoral geographers have largely concentrated their research efforts on partisan elections and voting systems, with comparatively little work on non-partisan votes including referenda and initiatives. Through study of referenda new insights on the juncture of context and politics can be made. As stated by Toal and Shelley (2003: 175):

> Analysis of the geographic distribution of initiatives, referenda, and other direct democratic processes has often proven a particularly valuable source of

> information to political geographers, because under direct democracy voters are expressing opinions on individual policy issues. Direct democracy provides especially valuable information about cultural and identity politics, and is therefore critical to the understanding of social and cultural linkages to political processes.

In the United States, for example, state legislatures can put referenda on the ballot for constitutional amendments in all states, and in two dozen states citizens may bypass the legislature and place changes to state statutes or amendments to state constitutions directly on the ballot. In 2010, 155 such measures were on the ballot in 36 different states. While 42 of these measures were placed on the ballot by citizens through the initiative process, 113 were placed on the ballot by state legislatures as referenda (National Conference of State Legislatures 2010). These measures pertained to a wide array of topics including taxes, state budgets, immigration, redistricting, abortion, stem cell research, marijuana and alcohol policy, among others. Such ballot measures ask citizens for a simple yes or no vote, and in conjunction with partisan elections, offer a more contextualized view of electoral geographies.

Over the past three decades "place" as an ontological unit tailored to electoral geography has also been extended (Agnew 1987, 1994, 1995, 1996, 1997; Agnew and Duncan 1989; Shin 2001; Shin and Agnew 2002, 2008). In these discussions, ballot preferences are attributed to the voter's social milieu rather than their demographic characteristics. Thus, voters from the Wire Grass region in southeastern Alabama may be considered an important contextual distinction in place rather than as simply a pocket of white middle class voters. Context, therefore, is engaged through the application of place as an ontological unit.

These developments in place-based electoral geography have largely been left unaddressed with quantitative methods, though recent publications indicate a shift in focus. Several authors (Chapman, Leib, and Webster 2007; Webster, Chapman, and Leib 2010; Webster and Leib 2001, 2002; Shin and Agnew 2002, 2008; Brown, Knopp, and Morrill 2005; Morrill, Knopp, and Brown 2007) demonstrate the move to a more contextualized quantitative electoral geography. These studies incorporate context inspired discussions of place-based electoral geographies alongside quantitative methods. Context, however, is often excluded from the quantitative models employed in these studies. The statistics in question have been used to extract spatial similarities associated with demographic profiles of selected electorates. In these methods place becomes an addendum to quantitative analysis rather than a fundamental component of it. The consequence is a normative stance in favor of quantitatively defined social similarity relative to place-based social difference, and the sacrifice of context as a central tenet within quantitative electoral geographies. This chapter seeks to address this bias with the incorporation of quantitatively defined place-based difference utilizing the local indicator of spatial autocorrelation (LISA) statistic. We follow the lead of Shin and Agnew (2002, 2008), Darmofal (2006), and (Seabrook 2009) in the use of this statistical approach in analysis of election returns.

The purpose of this chapter, then, is to extend the incorporation of context in electoral geography along three planes: the contrasts between partisan elections and referenda, in places, and with quantitative methodology. This chapter continues with a brief discussion of the 12 statewide votes examined, and is followed by a section detailing how we quantitatively defined Alabama's place-based electoral geographies. We follow this section with an overview of the political temperament associated with each place by relating their roles in the statewide votes. We conclude with a commentary relating what this effort adds to discussions regarding the nexus of context and quantitative methods as it applies to Alabama's electoral geographies.

Statewide Votes

We examined six statewide referenda and six statewide partisan elections in Alabama during the past dozen years. The referenda were legislatively placed on the statewide ballot as amendments to the state constitution between 1999 and 2006. They involved gambling, interracial marriage, tax reform, segregated schools, and the prohibition of same-sex marriage. All of these measures may be viewed as pertaining to culture war debates, at least from the lens of many in the Alabama electorate, with issues of morality being central to the decisions of many voters. The six recent statewide partisan elections included the presidential elections of 2000, 2004, and 2008, and the gubernatorial races of 1998, 2002 and 2006 (see Figure 11.1). A brief discussion of each of these statewide votes follows.

1998 Gubernatorial Election

The 1998 Alabama gubernatorial election pitted incumbent Republican Fob James against Democrat and sitting Lieutenant Governor Don Siegelman. James had won the 1994 gubernatorial election against the incumbent Democrat James Folsom, Jr. by less than 1 percent of the vote. Folsom's opposition to replacing the Confederate Battle flag on top of the state capitol played a role in James' narrow victory (Webster and Leib 2002). James' term had been controversial on several accounts, particularly in his treatment of education. He had also entered litigation arguing that student preachers had the right to engage in religious activity in the state's public schools, going so far as to submit a brief to the Supreme Court holding that only the federal government, and not the states, were bound by the Bill of Rights and the separation of church and state (Shelley and Webster 2000). While such efforts ingratiated James to the state's large fundamentalist religious community, they led to growing dissatisfaction in the business wing of the Republican Party.

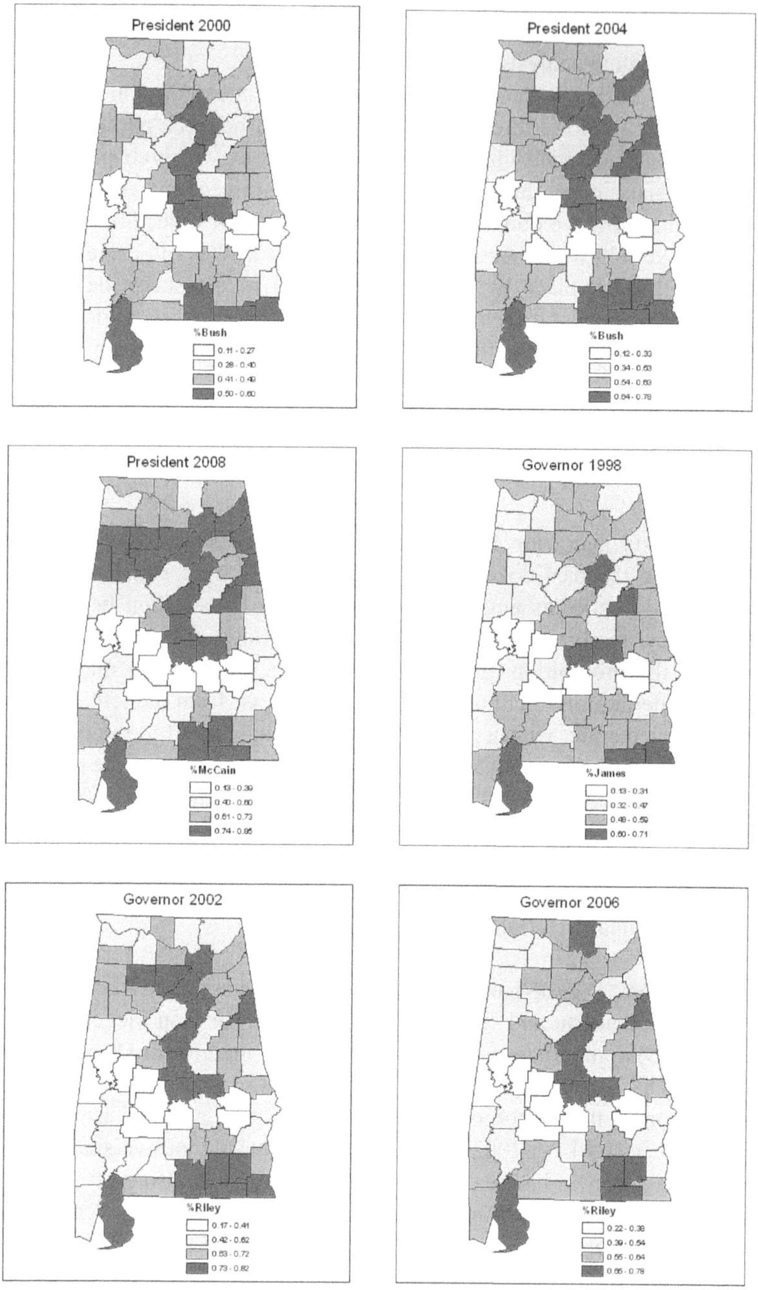

Figure 11.1 Partisan Elections

Siegelman's centerpiece campaign issue was the creation of an education lottery which used the proceeds to provide accomplished high school students with scholarships to Alabama colleges and universities. Dissatisfaction with James and interest in the lottery issue led Siegelman to victory with nearly 59 percent of the vote (Beyerle 1998b). Notably, there was record turnout in the election with 57 percent of registered voters casting ballots. Nearly 120,000 more Alabamians voted in the 1998 gubernatorial election than in 1994. Turnout in the state's predominately African-American counties was exceptionally high. In Sumter County which is 73 percent African-American, turnout was 74 percent. Similarly in Greene County which is 80 percent African-American, turnout was 78 percent (Beyerle 1998a).

2002 Gubernatorial Election

Democrat Siegelman ran for re-election in 2002. His Republican opponent was former Congressman Bob Riley. Polls prior to the November 5 election indicated a close contest. On election night Siegelman led and called on his opponent to concede. But after most returns were in, Republican stronghold Baldwin County reported a computer error and moved 7,000 votes from Siegelman to Riley. The result was that Siegelman fell behind by 3,117 votes out of the 1.3 million cast statewide, or 0.23 percent (Associated Press 2002). Alabama has no automatic recount statute, but Siegelman and his supporters pursued a statewide recount for two weeks without success. The Baldwin County district attorney, a Republican, "threatened to put everyone in jail" who tried to conduct a recount (Associated Press 2002). Finding few if any options to carry out a recount, Siegelman conceded on November 18 saying he did not wish to put the state through months of legal wrangling. As a result, Riley won the election with slightly over 50 percent of the vote.

2006 Gubernatorial Election

The 2006 gubernatorial election pitted incumbent Republican Riley against Democratic Lieutenant Governor Lucy Baxley. Both candidates faced notable opposition in the primaries with Riley winning the nomination against the former Chief Justice of the Alabama Supreme Court, and Baxley besting former Democratic Governor Don Siegelman. During the campaign Siegelman was being tried on racketeering charges making an active run all but impossible. Although Riley came under fire by his own party early in his first term due to his support for tax reform (Webster and Webster 2004), he slowly repaired his relationships with both social and business conservatives in the Alabama Republican Party. He further elevated his status due to his handling of the Hurricane Katrina relief effort in 2005. As a result, Riley easily won re-election with 55 percent of the vote.

2000 and 2004 Presidential Elections

During the past half century, Alabama has become a reliably Republican state in presidential elections with Georgian Jimmy Carter the only Democrat to win the state since John F. Kennedy in 1960. In 2000, George W. Bush won 58 percent of the state's vote, and increased his proportion to 63 percent in 2004. In 2000, Al Gore won only 17 of the state's 67 counties receiving large vote margins in the state's majority African-American counties. In Macon County, which is 85 percent African-American, Gore received over 87 percent of the vote. But in overwhelming white and suburban Shelby County, Gore received less than 22 percent of the 2000 presidential vote. In 2004, only 11 Alabama counties provided a majority of their votes for John Kerry, and all were in the Alabama Black Belt. Overwhelmingly African-American Macon County provided Kerry 83 percent of its vote, but overwhelmingly white Shelby County cast less than 20 percent of its ballots for Democrat Kerry. Thus, the geographic patterns of support and opposition to each Bush run were similar.

2008 Presidential Election

In the 2008 presidential election, Alabama again provided overwhelming support for the Republican candidate, John McCain. Although McCain garnered 45.6 percent of the nationwide vote, 60.3 percent of Alabama voters cast ballots for the Republican nominee. Barack Obama won 38.7 percent of the vote in Alabama in 2008, 2 percent better than John Kerry in 2004, but nearly 3 percent less than Al Gore in 2000. Obama won only 15 of Alabama's 67 counties with all but Jefferson County (Birmingham) being in the Alabama Black Belt – Kerry had won 11 counties in 2004, and Gore 17 in 2000. John McCain won 18 of the state's counties with more than 75 percent of the popular vote, with most of these counties being located in overwhelmingly white suburbs of Birmingham and the rural, principally white, counties in north central Alabama.

Exit polling after the 2008 presidential election indicated that Barack Obama won 43 percent of the white vote nationwide, equal to Bill Clinton's proportion in 1996, and the highest proportion for a Democratic nominee since 1976 (Edge 2010: 429). In comparison, Democrat Al Gore received 42 percent of the white vote in 2000, and John Kerry 41 percent in 2004 (Edge 2010: 430). In spite of Obama's comparatively successful results with white voters nationwide, he received support from only 10 percent of Alabama white voters in 2008, while garnering 98 percent support from the state's African-American voters (CNN 2008). White voter support in Alabama for Obama was the lowest in the country (Mississippi whites provided 11 percent of their support for Obama). In comparison, McCain received the support of 88 percent of white voters and only 2 percent support among black voters. Calling into question the pronouncements of some commentators about

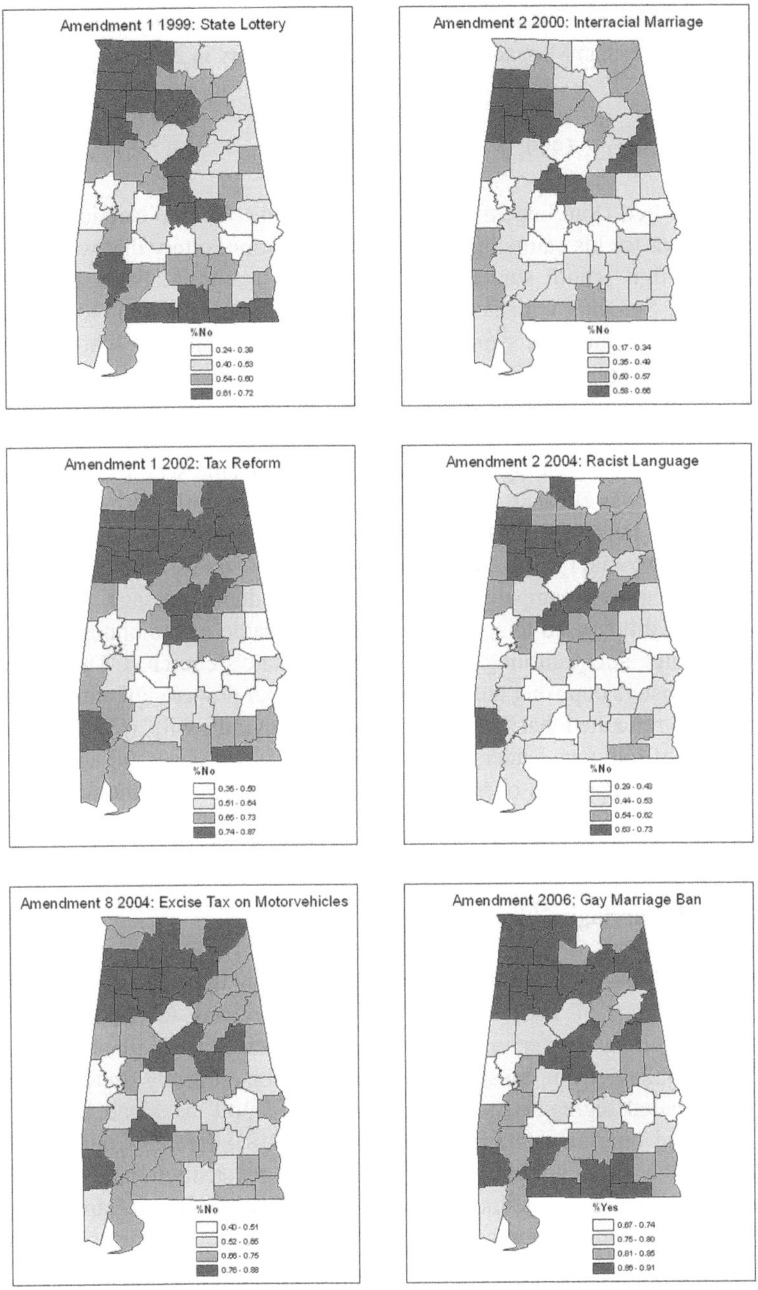

Figure 11.2 Amendments

the development of a "post-racial" electorate, Obama actually did worse among Alabama whites than Massachusetts Senator John Kerry in 2004 (Edge 2010: 437).

Education Lottery, 1999

As noted earlier, Don Siegelman won the 1998 gubernatorial election in part due to enthusiasm for his campaign to create an education lottery that would provide the state's high school students scholarships to attend Alabama colleges and universities. Students with B averages or above would be eligible for funding to attend any of the state's four year institutions, and all high school graduates would receive support to attend one of the state's community colleges regardless of their grade point averages. Proceeds from the lottery were also to be used for voluntary pre-kindergarten programs and to purchase computers for the public schools. Due to the funding problems faced by Alabama schools, initial support for the lottery was substantial. A poll in August 1999, only six week before the referenda was held, found 61 percent of the respondents were supportive of the proposal (DeMonia 1999). As the vote neared, the state's religious community mounted a vigorous campaign against the referendum's passage. The Alabama Christian Coalition used telephone messages and radio ads to oppose the lottery, and distributed anti-gambling pamphlets to the state's churches. The day before the vote ministers opposing the lottery held a 24-hour prayer vigil outside the state Capitol, and reiterated that "gambling goes against God's will" (CNN.com 1999). The efforts of religious leaders and organizations were successful and the lottery went down to defeat with over 54 percent of the Alabama electorate voting "no" on October 12, 1999.

Interracial Marriage, 2000

Alabama continues to operate under a constitution that was passed in 1901. The central goal of the century old constitution was to suppress African-American political power, and to insulate the state's large plantation owners and industrialists from taxation. The constitution severely limited property taxes, mandated segregated schools, implemented poll taxes and literacy tests, and outlawed interracial marriage. In 1967 the US Supreme Court found that prohibitions against interracial marriage were unconstitutional, making this section of Alabama's constitution a legal relict. In 1998, South Carolina deleted a similar section in its state constitution leaving Alabama as the sole remaining state with such a prohibition. In 1999, the Alabama legislature passed a bill to place a referendum on the ballot in 2000 that would strike the prohibition from the state's constitution. There was uniform support for passage of the referendum by the state's political leaders, and the only public opposition came from neo-Confederate groups. On November 8 the amendment passed with over 59 percent of the vote. But it is notable that a majority of voters in 25 of the state's 67 counties voted against excising the passage from the state's constitution. A Birmingham-Southern College political scientist estimated that only 44 percent of the state's white voters cast ballots in favor of the amendment (Webster and Quinton 2010).

Tax Reform, 2003

Upon his election to the governor's mansion in November 2002, Republican Bob Riley was faced with an extreme financial emergency in the state's budget. Riley, who was known as a strict fiscal conservative, proposed that a broad tax package to be submitted to the voters in September 2003. Alabama's tax system is one of the most antiquated in the United States, and its property tax rates are the lowest in the country. As a result, public education is funded with sales taxes which are famously unpredictable from year to year. Riley's tax reform package altered sales tax rates by eradicating the levy on food and drugs, reduced income taxes on low income earners while raising them modestly on upper income households, and raised property taxes on all categories of holdings. The plan would have raised over $200 million in new revenue, but it was estimated that as many as two-thirds of Alabama households would have experienced either no change or a reduction in their annual tax payments to the state (Webster and Webster 2004).

Republican responses to Riley's proposals were rapid and harsh, with the Alabama Republican Party Steering Committee voting overwhelmingly to oppose the Republican Governor's proposal. Farmers' groups, business groups, anti-tax groups and religious groups also strongly opposed the passage of the tax reform package. Notably, the National Christian Coalition came out in support of Riley because it reduced the regressive character of Alabama's tax system but the Alabama Christian Coalition vigorously opposed the package as a tax increase (Webster and Webster 2004). Education groups, African-American leaders and some business groups supported the tax reform package. On September 9, 2003, 68 percent of Alabama voters cast "no" votes on Riley's tax reform package. Only 13 of the state's 67 counties voted in favor of the amendment, with all being in or adjacent to the Alabama Black Belt.

Vote to Delete Racist Language from the State Constitution, 2004

Alabama continues to operate under its 1901 state constitution that was passed largely to limit the political power of African-American voters. The state's constitution included sections requiring poll taxes and segregated schools. The state constitution was also amended in 1956 in the wake of the *Brown vs. the Board of Education* decision stating that there was no "right" to a public education in Alabama at taxpayer expense. This amendment was passed to underscore the state's intention to shut down the public schools should integration be ordered by the federal courts. Due to these unenforceable sections in the state constitution, a bill was passed in the Alabama legislature to delete such passages through a referenda held in November 2004.

The primary opposition to passage came from several of the state's conservative religious leaders. They argued that deleting the passage that there is no right to a public education in Alabama would result in lawsuits to increase taxes to more fully support the state's schools. The President of the Alabama Christian Coalition went so far as to characterize education as a "gift" from the state as opposed to a "right". Notably, litigation in the early 1990s had already determined that there is a right to a public education in Alabama (Shah 2005).

In November 2004 the amendment was rejected by less than 2,000 votes out of the 1.4 million cast, or by 0.13 percent. Less than half of the state's 67 counties supported passage of the amendment. These counties included those with major cities and those with large African-American populations. The largest cluster of counties rejecting passage was located in the northern part of the state. The population in this part of Alabama is overwhelmingly white and includes large proportions of religious conservatives (Webster and Quinton 2010).

"Motor Vehicle Taxes Act," 2004

The "Motor Vehicle Taxes Act" would have amended Amendment 93 of the state Constitution of 1901 to authorize the state legislature to repeal state property taxes on vehicles over 26,000 pounds in weight, except those used for farm and forest production. In lieu of the *ad valorem* tax the legislature could instead levy an excise tax on such vehicles. The Alabama state constitution requires that motor vehicle excise taxes be spent on public highways, and *ad valorem* taxes be allocated to the local governments in the jurisdiction in which they are collected. Amendment 8 would have created an exception for excise taxes collected on vehicles over 26,000 pounds to allow them to replace the *ad valorem* taxes lost as a result of its passage. The excise taxes were to be apportioned based on miles traveled inside and outside the state of Alabama.

Amendment 8 received only limited attention by the news media. Arguably, its purpose was confusing to many. A tax lobby group reported that the Alabama Department of Revenue and Alabama Trucking Association supported the Amendment (Arant 2004). In opposition, the Conservative Christians of Alabama recommended a no vote, arguing that if Amendment 8 was passed the legislature would "have the authority to set the rate of collection and direct how this money is spent. In times past, this has been called a slush fund" (Conservative Christians of Alabama 2004).

Constitutional amendments in Alabama pertaining to taxes generally face strong opposition, with many voters assuming that any change will result in a tax increase (Webster and Webster 2004). Amendment 8 was defeated with 68.5 percent of the voters casting "no" votes. Only two counties voted in favor of the Amendment – Greene and Sumter – both in the Alabama Black Belt. Other surrounding counties with large African-American populations rejected the Amendment, but with much closer results than the state as a whole. The counties most in opposition were generally located in the northern portion of the state. Nearly 86 percent of Limestone County and 83 percent of Fayette County voters cast ballots in opposition.

Same-Sex Marriage Vote, 2006

The final referendum examined pertains to a constitutional prohibition against same-sex marriage. This referendum is one of over two dozen to have been held since November 2003 when the Massachusetts Supreme Court ruled that barring same-sex marriage was unconstitutional. It is also notable that many of these states, including Alabama, already had statutes defining marriage in exclusively heterosexual terms. Although many states in the South are less than receptive to gay and lesbian issues, Alabama has a particularly vicious record. In July, 2004 alone, two gay men were murdered because of their homosexuality. As a result, Alabama has been labeled the "worst place to be a gay person" in the United States, and the "homophobic center of the universe" (Webster, Chapman and Leib 2010).

The Alabama legislature passed a bill to place an anti-same sex marriage referendum on the ballot in 2005. Support for the bill was overwhelming, with only limited opposition from a handful of African-American legislators. Because there was little doubt about the vote's outcome, there was only limited campaigning for and against the amendment. The most visible support for the measure came from the Alabama Christian Coalition which distributed 1.2 million voting guides around the state urging support for the amendment. The limited opposition came from the ACLU and Equality Alabama, the latter being a gay rights group. The amendment passed in every Alabama county, garnering 81 percent of the statewide vote. The counties with the lowest levels of support were found in the Alabama Black Belt, and the amendment found its greatest support in the northern part of the state. In both Marion and Winston counties, for example, voters cast over 90 percent of the support for the amendment.

Alabama's Place-Based Electoral Geographies Quantitatively Defining Context: LISA Analysis

Local indicators of spatial autocorrelation (LISA) were used to delineate the geographies associated with our suite of statewide returns. LISA statistics were calculated with the univariate LISA option in GeoDa. In this case the features identified as constituent parts of LISA clusters were Alabama's counties. Though we do not assume that county electorates in Alabama are homogeneous for the purposes of this study, county level vote returns provide an appropriate scale for analysis of statewide votes. The LISA statistic indicated clusters of counties in terms of support (high-high) or lack of support (low-low) for each Republican candidate in the partisan elections and the conservative position for each referendum. The LISA statistic also identified counties that are distinct outliers from local patterns. These outliers were singled out as either highly supportive of the Republican candidate or conservative position in an area of tepid support (high-low), or less-supportive of the Republican candidate or conservative position in an area of intense support (low-high) (Anselin 2005).

The calculations yielded LISA clusters that were statistically significant at .05. Anselin (2005) notes, however, that LISA clusters often extend beyond the statistically significant features produced by GeoDa into surrounding features. Thus, we chose to forego the simplistic threshold of statistical significance produced in GeoDa with an alternative cartographic representation of statistical significance. To complete this task we assigned LISA cluster values to each county regardless of statistical significance utilizing the Moran scatter plot utility in GeoDa. This data was transferred into ArcGIS where the dot density tool was utilized to display the significance measures as an element of the cartographic presentation. Each dot represents .001 of the total significance as measured by a P-value. As the P-value increases and the LISA value becomes less statistically significant, the number of dots in the county increases. As a result, counties with the most statistically significant values have few dots, while those with less statistically significant values have high dot densities. In this way, we moved beyond cartographic representation of statistical significance with a threshold of .05 or better to a representation of continuous statistical significance. This allowed us to acknowledge that highly significant features are only the core of larger LISA clusters.

The results of the LISA analysis are included in Figures 11.3 (partisan elections) and 11.4 (referenda). There are reoccurring spatial patterns associated with the returns from both the partisan elections and the referenda. There are also spatial consistencies related to support for Republican candidates that are distinct from those found relating to the conservative positions in six referenda. In particular the Black Belt stood out in its resistance to both Republicans and the conservative position expressed in the referenda. Alternatively, the partisan elections and referenda had different areas of high support, the Wiregrass and north Alabama, respectively. Further, Jefferson, Madison, and Lee counties all stood out as consistent outliers from local patterns. We hypothesize these spatial cleavages are the durable underlying electoral geographies associated with the varied contexts of recent statewide elections in Alabama.

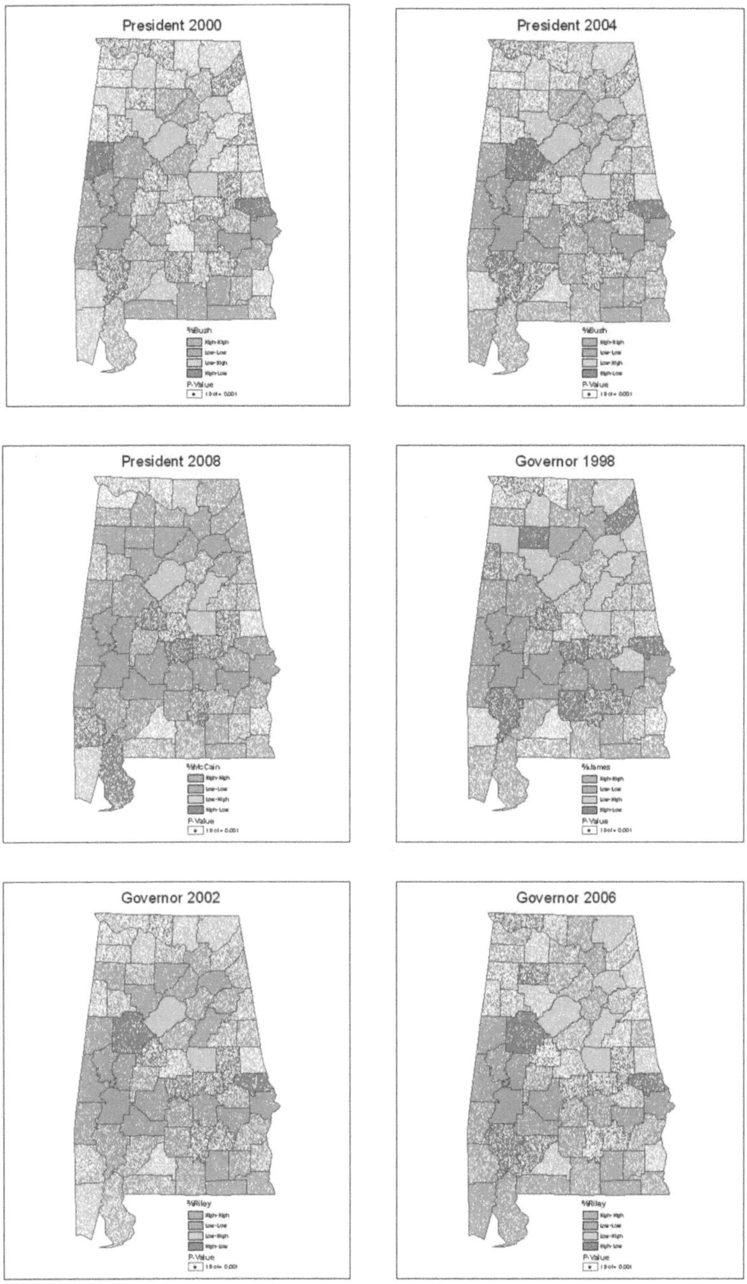

Figure 11.3 Results of the LISA Analysis for Partisan Elections

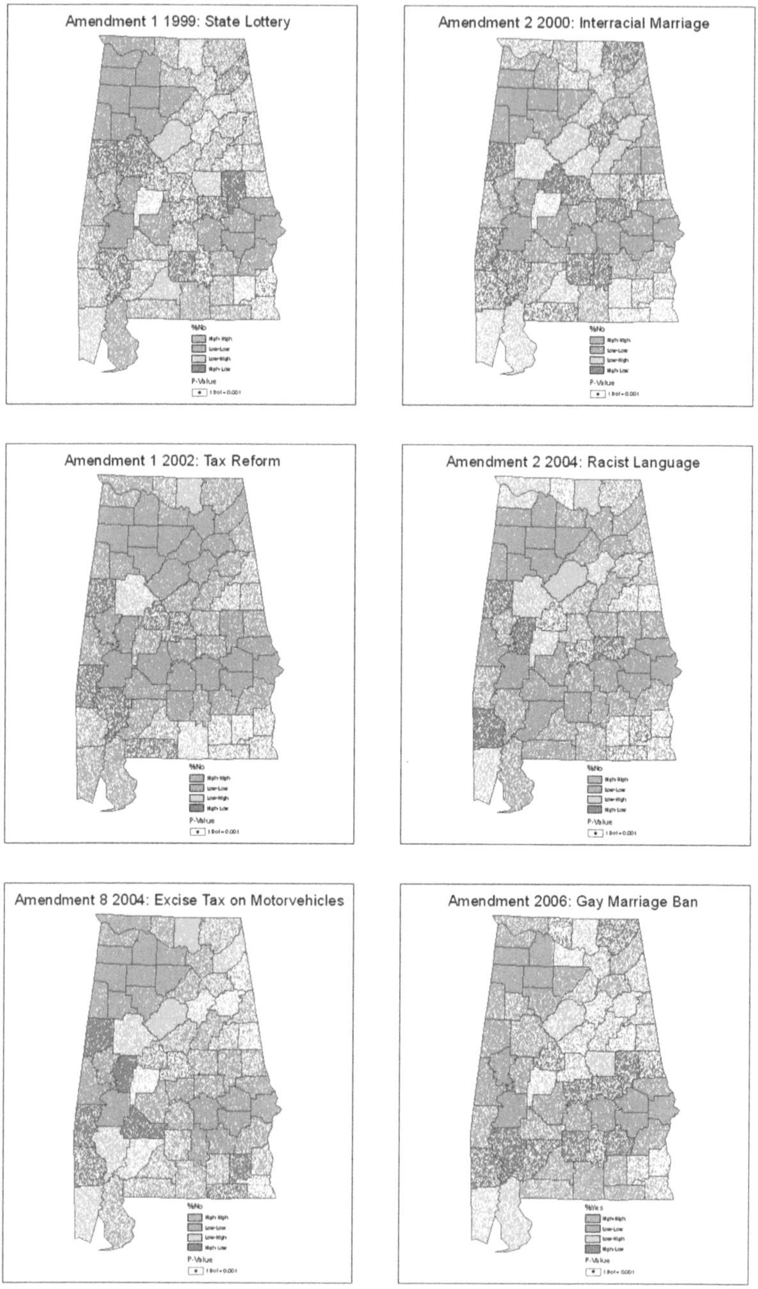

Figure 11.4 Results of the LISA Analysis for Referenda

Table 11.1 Principal Axis Factor Analysis with an Oblique Rotation (N=67)

Variable	Factor I Loadings	Factor II Loadings	Extraction Communalities
Presidential Election 2000, % Bush	.952	.626	.906
Presidential Election 2004, % Bush	.995	.703	.994
Presidential Election 2008, % McCain	.905	.862	.935
Gubernatorial Election 1998, % James	.974	.563	.964
Gubernatorial Election 2002, % Riley	.976	.783	.980
Gubernatorial Election 2006 % Riley,	.949	.556	.913
Education Lottery 1999, % No	.813	.832	.809
Delete Interracial Marriage Ban 2000, % No	.580	.872	.761
Tax Reform, % No	.677	.897	.814
Delete Racist Language 2004, % No	.549	.934	.885
Truck Excise Tax 2004, % No	.487	.868	.772
Ban Same-sex Marriage 2006, % Yes	.711	.876	.794
Extraction Sum of Squared Loadings	8.993	1.533	
% of Variance	74.938	12.774	
Cumulative % of Variance	74.938	87.711	

Quantitatively Gauging Spatial Alignment: Factor Analysis

Given the clear geographic patterns of the LISA clusters we employed a factor analysis as another test for the presence of durable geographies amongst Alabama's statewide elections (Table 11.1). We utilized an oblique rotation since there was no expectation that two uncorrelated factors would be realistic given the similarities in the spatial extents of the LISA clusters for the two groups of elections. Two factors with eigenvalues above 1.0 were generated accounting for 89.7 percent of the variance. The first factor accounted for 75.9 percent of the common variance and was primarily associated with partisan elections, with all six having loadings exceeding .90. Notably, the lowest loading for the six elections was for the 2008 presidential election at .905, while the remaining five partisan contests had loadings ranging from .949 (2006 gubernatorial election) to .995 (2004 presidential election). All six referenda had significant but lower loadings on Factor I ranging from .487 (truck excise tax) to .813 (education lottery). Thus, among the six referenda, the lottery vote was most similar to the partisan elections.

Factor II accounted for 12.8 percent of the common variance, with generally higher loadings by the six referenda ranging from .832 to .934. Not surprisingly given its alignment with partisan election on Factor I, the lowest loading by the six referenda was yielded by the education lottery vote. The six partisan elections had lower loadings on Factor II with the exception of the 2008 presidential election. Although the other partisan contests had loadings on Factor II ranging from .556 (1998 gubernatorial election) to .783 (2002 gubernatorial election), the 2008 presidential election yielded a loading of .862, higher than two of the loadings for the six referenda on Factor II. Thus, the 2008 election had the characteristics of both a partisan contest and a culture war related referenda. Arguably, race played a role since only one in ten white Alabama voters cast their ballots in support of Barack Obama.

The two factors were inter-correlated at .67, indicating a high level of consistency not only within the referenda and partisan elections as distinct categorizations, but also between the two sets of statewide votes. Two statewide votes demonstrate this relationship: the lottery referenda from 1999 and the presidential election of 2008. The percent of "no" votes for the lottery amendment produced high loadings on both Factor I (.813) and Factor II (.832). The percent of votes cast for John McCain in 2008 also yielded high loadings on both Factor I (.905) and Factor II (.862). The 1999 lottery referendum and 2008 presidential election returns yielded similar geographic patterns, which included a persistent effect from the Black Belt and a general agreement between north Alabama and the Wiregrass. These two statewide votes were the only contests in which both north Alabama and the Wiregrass returned strong support for the same issue or candidate. Thus the durability of the Black Belt coupled with the shifting importance of the Wiregrass and north Alabama in the statewide votes contributed to the high level of concurrent similarity and difference identified in the factor analysis. These consistent differences and similarities within the returns from both the referenda and partisan elections further confirmed our conclusions from the LISA analysis.

Discussion of Results

Even though the context of each statewide vote was different, coherent spatial cleavages exist amongst the election returns analyzed. We attribute this spatial durability to identifiable differences associated with different places in Alabama – place as the site where people live their lives and where their views of the world are shaped (Agnew 1987, 1994, 1995, 1996, 1997; Agnew and Duncan 1989; Shin and Agnew 2002, 2008). These places are enlivened by the context of peoples' daily routines and reflect collective decision making in regards to the contexts for each statewide election. Though the entirety of the electorates in each place most certainly did not vote for the same candidates or issues, enough did so to identify distinct spatial differences in Alabama's electoral geographies. These places included a major urban locale (Birmingham), the Black Belt, the Wiregrass, counties with concentrations of creative class employment (Madison and Lee

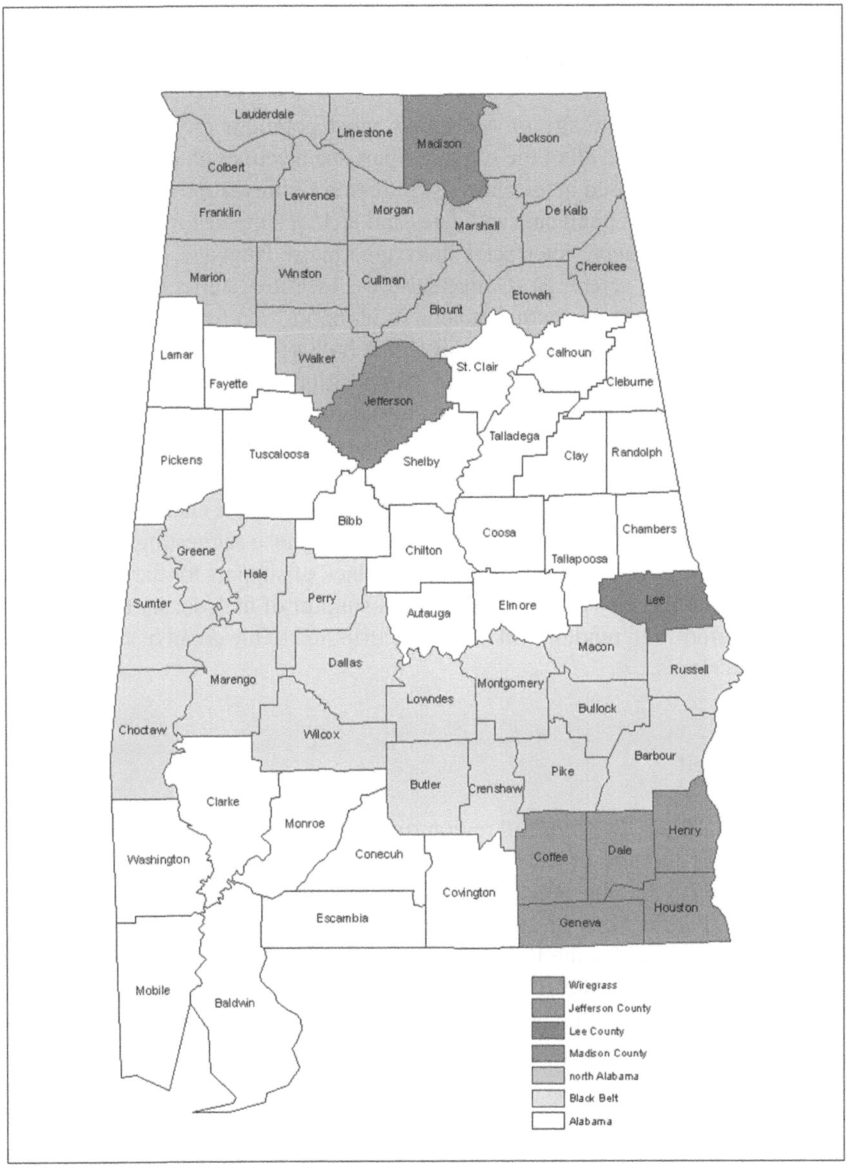

Figure 11.5 Important Places in Alabama

counties), and the counties in the northern third of Alabama (Figure 11.5). Each of these places has a distinct electoral character associated with past levels of support in statewide votes (Webster 1996; Webster and Quinton 2010; Webster and Leib 2002; Webster and Webster 2004). The roles played by each, whether in support or opposition, are discussed below.

Jefferson County – Birmingham

Jefferson County is the site of Alabama's most populous city, Birmingham. It played a distinct role in all of the statewide partisan elections analyzed, and in four of the referenda included in the analysis. Jefferson County stood out in terms of its lack of support for Republican candidates and lack of support for the conservative positions in the lottery, interracial marriage, racist language, excise tax, and same sex marriage referenda. Conversely, we identified Jefferson County as a significant addition to northern Alabama in its opposition to tax reform in 2003. While conceivably out of character given the bulk of the statewide elections we analyzed, we attribute Jefferson County's opposition to tax reform to the creative class (Florida 2004). Thirty-two percent of the county's workforce are employed in creative class occupations. Members of the creative class generally hold more liberal views than the population as a whole (Chapman, Leib, and Webster 2007). However, Webster and Quinton (2010) found that the creative class in Alabama can be influenced by economic arguments, particularly those suggesting higher taxes. Thus, Jefferson County's vacillation between lack of support for the conservative position in the majority of the referenda and support of the conservative position in the tax reform referendum can likely be attributed to the creative class concern over tax increases.

The Black Belt

The Alabama Black Belt is a soil region running across the lower half of the state that served as the site of the state's Antebellum plantation economy (Webster and Samson 1992). The Black Belt was predominantly populated by African-Americans (Figure 11.6) in 2000, and it is largely rural. This region provided very limited support for the Republican candidate in all of the partisan elections examined. In so doing the Black Belt continued its longstanding support of the Democratic Party (Webster and Bowman 2008). The Black Belt also opposed the conservative position in all six of the referenda analyzed. While undoubtedly some African-American residents of the Black Belt have supported the conservative position on culture wars referenda in Alabama (Webster, Chapman, and Leib 2010), the region as a whole was demonstrably less supportive of the conservative positions in these referenda than the state as a whole.

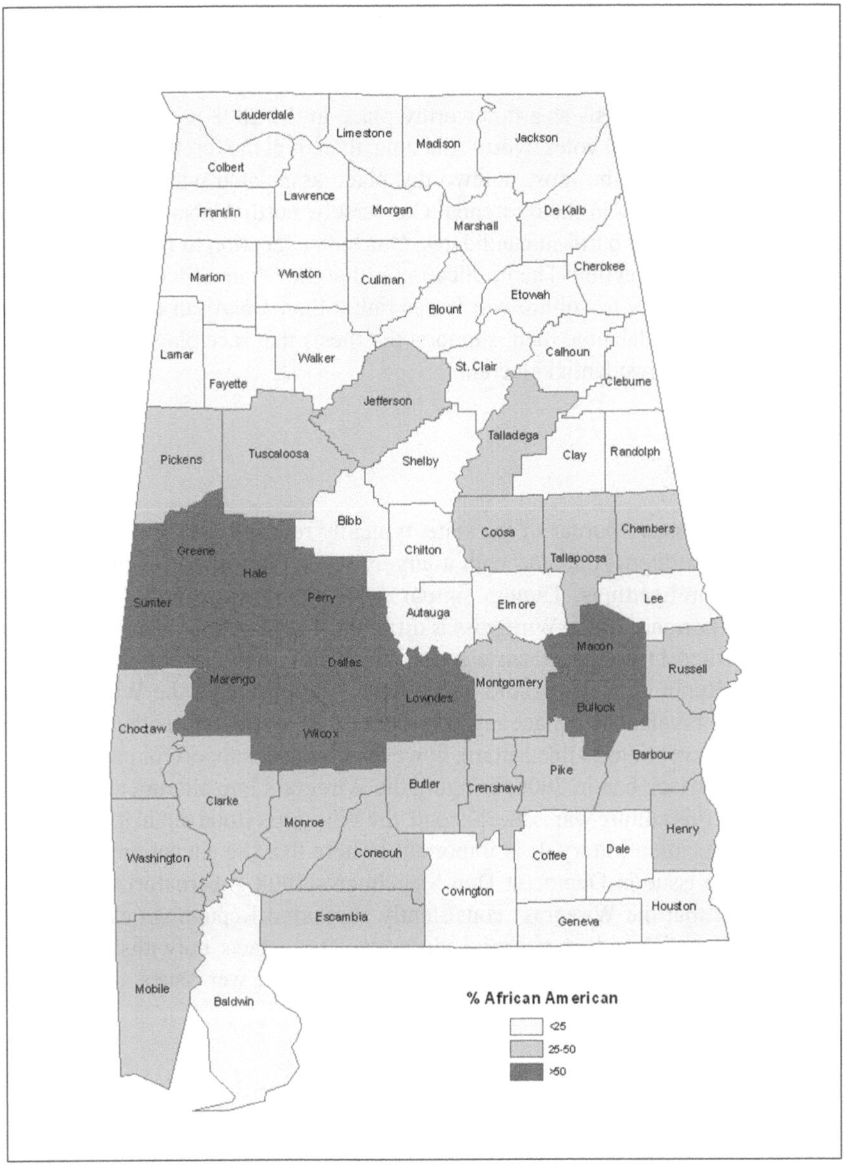

Figure 11.6 Percentage of African-Americans in Alabama by County

North Alabama

Race continues to be a major fault line in Alabama's electoral geographies with white and black voters having great contrasts in their voting behaviors. Strongly associated with this division is religious conservatism, which characterizes many white rural places in the state (Webster and Webster 2004). The counties located in the northern third of Alabama are both substantially rural, largely white and stood out in our LISA analysis as a noteworthy place in the 2008 presidential election and all of the referenda votes. North Alabama, then, is at the forefront of the culture wars in the state as the most noteworthy place associated with support for the conservative position in the referenda. Conversely, north Alabama rarely showed coherent support for partisan candidates. Our lone exception to this trend was the 2008 presidential election. The implication is that the Obama/McCain contest was interpreted similarly to culture war issues rather than a partisan election in north Alabama. North Alabama, then, supports the thesis that race played a substantial role in the 2008 presidential election.

The "Wiregrass"

The far southeastern corner of the state, typically referred to as the "Wiregrass," is similar to northern Alabama with a largely white population residing in small towns and rural settings. Though similar in many ways to north Alabama, the electoral importance of the Wiregrass is different. The Wiregrass was an important place identified in each partisan election. As such it was the only continuously significant Republican stronghold in the state. With that said, the Wiregrass was relatively ambivalent as a place regarding almost all of the referenda. There were two notable exceptions to this pattern, however. First was support for passage of the same-sex marriage ban in 2006, bringing the Wiregrass into alignment with north Alabama on this culture war issue. Second, the Wiregrass stood out in its opposition to a state education lottery. It is important to note that the education lottery was the signature issue in Democrat Don Siegelman's 1998 gubernatorial campaign, and the fact that the Wiregrass consistently supported Republican candidates in both gubernatorial and presidential elections. These facts notwithstanding, the Wiregrass did align with north Alabama on these culture war issues.

Madison and Lee Counties

Recently, the creative class (Florida 2004) has become a part of electoral research in the broader South (Chapman, Leib, and Webster 2007) including Alabama (Webster, Chapman, and Leib 2010; Webster and Quinton 2010). Two counties consistently stood out as distinct outliers from broader regional vote returns due to concentrations of the creative class employment: Lee and Madison. Lee County is the site of Auburn University and is located on the edge of the Black Belt. Madison County is the location of the University of Alabama-Huntsville, and is also distinct in its employment base which is associated with federally funded rocket science (Rogers et al. 1994).

In terms of partisanship, Madison County was a significant outlier in the 2008 presidential election as a place offering relatively limited support for Senator John McCain. Alternatively, in both the 1998 and 2006 gubernatorial elections, Madison County was part of a small cluster of counties that provided support for the Republican nominees. Further, Madison County was identified as a regional outlier for its lack of support for the conservative position in the tax reform, the deletion of racist language from the state constitution, and the large truck excise tax referenda. These results reveal little coherence regarding Madison County's political temperament. We can, however, definitively conclude that Madison County is distinct from the rest of north Alabama, and indeed is distinct from much of Alabama more broadly speaking.

Lee County was identified as a noteworthy outlier in five of the partisan contests: the 2000 and 2004 presidential contests and the 1998, 2002, and 2006 gubernatorial elections. In each of these contests, Lee County was significant in its support for the Republican candidate. Interestingly, in 2008 Lee County was also a significant place, though in this case its vote returns were similar to the Black Belt in its lack of support for Senator McCain. When considering the six referenda, Lee County was noteworthy because of its lack of support for the conservative position. In these cases its vote returns were similar to those of the Black Belt. Thus while Lee County offers more support for Republicans than the surrounding counties of the Black Belt, it aligns with the Black Belt on the culture wars.

Conclusions

This analysis identified durable geographies associated with both the referenda and partisan elections. These included strong support for Republican candidates in the Wiregrass and consistent support for the conservative position in the referenda in north Alabama. Further, Jefferson, Madison, and Lee counties were all consistently distinct from broader regional vote patterns. These differences were intertwined with strong similarities in Alabama's electoral geographies evident throughout all of the elections. In particular the Black Belt consistently opposed both Republican candidates and the conservative positions on all six of the referenda. Thus, there were patterned spatial alignments associated with both the partisan elections and referenda.

With that said, these spatial patterns changed depending on both the context of the election and that of the place. Vote returns from the Wiregrass, north Alabama, Jefferson, Madison, and Lee counties all reflect contingent outcomes dependent upon context. Indeed, only the Black Belt demonstrated firm durability throughout all of the elections analyzed. But even in this case, if we go far enough back in time we can see electoral change in the Black Belt (Webster 1992). Contingency, then, is the dominant shape of electoral outcomes in Alabama. In this sense a focus on context in place and associated with specific elections is necessary for a thorough understanding of electoral processes and geographies in Alabama. Within this process quantitatively defined place-based electoral cleavages have proven to be helpful tools in accomplishing this task. We finish then with a reaffirmation of the engagements already conducted with both context and quantitative methods in electoral geography and forward the call for the use of context sensitive quantitative methods.

References

Agnew, J. 1987. *Place and Politics: The Geographical Mediation of State and Society*. Boston: Allen & Unwin.

Agnew, J. 1994. "The national versus the contextual – The controversy over measuring electoral change in Italy using Goodman flow-of-vote estimates." *Political Geography* 13: 245–54.

Agnew, J. 1995. "The rhetoric of regionalism – The Northern League in Italian politics, 1983–94." *Transactions of the Institute of British Geographers* 20: 156–72.

Agnew, J. 1996. "Mapping politics: How context counts in electoral geography." *Political Geography* 15:129–46.

Agnew, J. 1997. "The dramaturgy of horizons: Geographical scale in the 'reconstruction of Italy' by the new Italian political parties, 1992–95." *Political Geography* 16: 99–121.

Agnew, J. and J. Duncan. 1989. *The Power of Place: Bringing Together Geographical and Sociological Imaginations*. Boston: Unwin Hyman.

Anselin, L. 2005. "Exploring Spatial Data with GeoDa: A Workbook." Urbana: University of Illinois, Spatial Analysis Laboratory, Department of Geography.

Arant, B. 2004. "An overview of the proposed Constitution amendments affecting Alabama taxes or economic development." *State and Local Tax Bulletin*, 28 October.

Associated Press 2002. "Black turnout made election close, NAACP says," *Birmingham News*, 9 December: 4B.

Barnes, T. 2009. "'Not only … but also': Quantitative and critical geography." *Professional Geographer* 61: 292–300.

Beyerle, D. 1998a. "Election turnout called highest in history," *Tuscaloosa News*, 19 November: 1B, 3B.

Beyerle, D. 1998b. "GOP Right says voters anti-fob, not pro-lottery," *Tuscaloosa News*, 2 December: 2B.

Brown, M., L. Knopp, and R. Morrill. 2005. "The culture wars and urban electoral politics: Sexuality, race, and class in Tacoma, Washington." *Political Geography* 24: 267–91.

Chapman, T., J. Leib, and G. Webster. 2007. "Race, the creative class, and political geographies of same sex marriage in Georgia." *Southeastern Geographer* 47: 27–54.

Conservative Christians of Alabama 2004. "November 2nd vote on Alabama Constitutional Amendments." Available at: http://ccofal.org/alabama/2004/amendment_recommendations.phtml [last accessed: March 31, 2009].

CNN.com 1998. "Alabama voters to decide on lottery," 12 October. Available at: http://www.cnn.com/ALLPOLITICS/stories/1999/10/12/bama.lottery/ [last accessed: December 15, 2003].

CNN.com. 2008. "Local Exit Polls – Election Center 2008." Available at: http://www.cnn.com/ELECTION/2008/results/polls [last accessed: September 27, 2010].

Darmofal, D. 2006. "The political geography of macro-level turnout in American political development," *Political Geography* 25: 123–50.

DeMonia, R. 1999. "Lottery gets strong support." *Birmingham News* 29 August: 1A, 8A.

Edge, T. 2010. "Southern strategy 2.0: Conservatives, white voters and the election of Barack Obama," *Journal of Black Studies*, 40(3): 426–444.

Florida, R. 2004. *The Rise of the Creative Class: And How it's Transforming Work, Leisure, Community and Everyday Life*. New York, NY: Basic Books.

Morrill, R., L. Knopp, and M. Brown. 2007. "Anomalies in red and blue: Exceptionalism in American electoral geography." *Political Geography* 26: 525–53.

National Conference of State Legislatures 2010. "2010 Ballot Measures Preview," 22 September. Available at: http://www.ncsl.org/default.aspx?tabid=20114.

Rogers, W., R. Ward, L. Atkins, and W. Flynt 1994. *Alabama: The History of a Deep South State*. Tuscaloosa, AL: University of Alabama Press.

Seabrook, N. 2009. "The Obama effect: Patterns of geographic clustering in the 2004 and 2008 presidential elections." Forum-a Journal of Applied Research in Contemporary Politics 7: Article 6

Shah, P. 2005. "The hypothetical impact of Alabama's failed Amendment 2 on public school funding," *Alabama Law Review* 56: 863–72.

Shelley, F. and G. Webster 1998. "Population, settlement, race, and ethnicity in the South." *Journal of Geography* 97(4/5): 163–76.

Shin, M. 2001. "The politicization of place in Italy." *Political Geography* 20: 331–52.

Shin, M. and J. Agnew. 2002. "The geography of party replacement in Italy, 1987–1996." *Political Geography* 21: 221–42.

Shin, M. and J. Agnew. 2008. *Berlusconi's Italy: Mapping Contemporary Italian Politics*. Philadelphia, PA: Temple University Press.

Toal, G. and F. Shelley 2003. "Political geography," in Gary L. Gaile and Cort J. Willmott (eds), *Geography in America at the Dawn of the 21st Century*. Oxford: Oxford University Press, 164–84.

Warf, B. and S. Arias. 2008. *The Spatial Turn: Interdisciplinary Perspectives*. London; New York: Routledge.

Webster, G. 1993. "Congressional redistricting and African-American representation in the 1990s – An example from Alabama." *Political Geography* 12: 549–64.

Webster, G. 1996. "Partisan shifts in presidential and gubernatorial elections in Alabama, 1932–1994." *Professional Geographer* 48: 379–91.

Webster, G. and J. Bowman. 2008. "Quantitatively delineating the Black Belt geographic region." *Southeastern Geographer* 48: 3–18.

Webster, G., T. Chapman and J. Leib. 2010. "Sustaining the 'societal and scriptural fence': Cultural, social, and political topographies of same-sex marriage in Alabama." *Professional Geographer* 62: 211–29.

Webster, G. and J. Leib. 2001. "Whose South is it anyway? Race and the Confederate battle flag in South Carolina." *Political Geography* 20: 271–99.

Webster, G. and J. Leib. 2002. "Political culture, religion, and the Confederate battle flag debate in Alabama." *Journal of Cultural Geography* 20: 1–26.

Webster, G. and N. Quinton. 2010. "The electoral geographies of two segregationist ("Jim Crow") referenda in Alabama." *Political Geography* 29: 370–80.

Webster, G. and S. Samson. 1992. "On defining the Alabama Black Belt: Historical changes and variations." *Southeastern Geographer* 32: 163–72.

Webster, G. and R. Webster. 2004. "Taxing issues: Geography, politics and the 2003 Tax Reform Referendum in Alabama." *Southeastern Geographer* 44: 190–215.

Chapter 12
Sexual Politics and the Sunshine State: Electoral Geographies of Florida's Marriage Protection Amendment

Thomas Chapman

In 2008, Florida voters approved the Florida Marriage Protection Amendment,[1] also known as *Proposition 2*, by a margin of 62.1 percent to 37.9 percent, making Florida the 27th state to prohibit legal recognition of same-sex marriage by amending its state constitution. This chapter explores the electoral geographies of *Proposition 2* by examining the spatial distribution of the vote through a lens of the cultural politics of sexuality. I begin with a brief overview of the scalar implications of the debate on same sex marriage in the United Sates, followed by a discussion of Florida's "place" in the national debate. I then examine the socio-cultural implications of the debate that lead up to the actual vote, taking a cue from the recent subgenre of work that invokes a critical and post-positivist approach towards the cultural politics of sexuality and electoral outcomes (see, for example Brown, Knopp, and Morrill, 2005; Rasmussen, 2006; Webster, Chapman, and Leib, 2010). I conclude with a discussion of the spatial outcome of the vote, revealing a multiplicity of cultural, social, and political geographies at the local scale.

The Scalar Politics of Same-Sex Marriage Referenda

In 1957, Wyoming was the first state to insert a statutory prohibition on same-sex marriage, followed by Maryland in 1973 and New Hampshire in 1983. But the notion of extending marriage rights to same-sex couples in the United States did not become a significant political issue until the mid-1990s, when the Hawaii Supreme Court struck down that state's constitutional ban on same sex marriage.[2] By most measures, the Hawaii case was the springboard that set the stage for a national debate on the issue, reacting to fears that the Hawaii case would spur

1 The official ballot language read as follows: "Marriage as the legal union of only one man and one woman as husband and wife. No other legal union that is treated as marriage or the substantial equivalent thereof shall be valid or recognized."

2 *Baehr v. Lewin* 74 Haw. 530; 852 P.2d 44 (1993).

legal challenges elsewhere. Indeed, the Hawaii decision motivated few state legislatures to pass their own Defense of Marriage statutes. At the federal level, Congress passed, and President Clinton signed into law the Defense of Marriage Act (DOMA)[3] in 1996, limiting the definition of legal marriage to opposite sex couples (for federal purposes), and allowing individual states to reject recognition of same sex marriages performed in other states. But by all accounts, the 2003 landmark decision by the Massachusetts Supreme Court granting that state's same-sex couples the legal right to marry set the stage for a flurry of legislative action to pass constitutional amendments banning same-sex marriage (Table 12.1). Several states, including Florida, had existing statutory laws banning same-sex marriage on the books. But those most opposed to granting marriage rights to same-sex couples argued that a constitutional amendment was necessary in order to provide a stronger legal remedy that could withstand future judicial scrutiny (Webster, Chapman and Leib, 2010). The national debate reached a fever pitch in 2004, when President Bush proposed an amendment the US Constitution that would limit legal marriage to opposite-sex couples. The notion of a federal

Table 12.1 Defense of Marriage Legislative and Court Actions*

2003	• Massachusetts Supreme Judicial Court rules state constitution guarantees equal marriage rights for same sex couples.
2004	• President Bush announces support for federal constitutional amendment. • Same sex couples begin marrying in Massachusetts. • US Senate votes to block constitutional amendment to ban same sex marriage. • US House of Representatives votes against federal constitutional amendment banning same sex marriage. • Arkansas, Georgia, Kentucky, Louisiana, Missouri, Michigan, Mississippi, Montana, North Dakota, Ohio, Oklahoma, Oregon and Utah voters approve constitutional amendments banning same sex marriage in those states.
2005	• New York Supreme Court justice rules state law prohibiting same sex marriage is unconstitutional. • California Superior Court judge rules ban on same sex marriage is unconstitutional. • Kansas voters approve constitutional amendment banning same sex marriage. • Navajo Nation Tribal Council votes to ban same sex marriage; Navajo President subsequently vetoes the measure to ban same sex marriage. • California legislature becomes first legislative body in US to pass a bill legalizing same-sex marriage without a court order. Governor Arnold Schwarzenegger vetoes the legislation. • Texas approves a constitutional amendment prohibiting same-sex marriage. • Federal Court strikes down Nebraska's same-sex marriage ban in the state constitution.

3 Pub. L. 104-199, 100 Stat. 2419 (Sept. 21, 1996). See http://www.domawatch.org.

Table 12.1 Continued

2006	• Alabama passes a constitutional amendment banning same sex marriage. • Supporters of a constitutional ban on same sex marriage in Florida fail to collect enough signatures for the issue to appear on that year's general election ballot. • The New Jersey Supreme Court orders the legislature to redefine marriage to include same-sex couples or to establish a separate legal structure to provide same-sex couples the same benefits as opposite sex couples. • South Dakota, South Carolina, Virginia, Wisconsin, Tennessee, Colorado and Idaho pass constitutional amendments defining marriage between a man and a woman. • Arizona becomes the only state in which a constitutional amendment to define marriage between a man and woman fails.
2007	• Massachusetts legislators pass legislation preventing a proposed constitutional amendment banning same-sex marriage from being placed on the ballot. • The Maryland Supreme Court overturns a lower court decision ruling that same-sex couples do not have the right to marry.
2008	• The California Supreme Court rules that banning same-sex marriage is unconstitutional. • Florida, Arizona, and California voters pass constitutional amendments defining marriage between a man and a woman.
2009	• The Iowa Supreme Court rules that banning same-sex marriage is unconstitutional. The ruling takes effect immediately. • The Vermont legislature passes a statute allowing same-sex marriage. Vermont is the first state in which legislature, not the courts, authorized same-sex marriage.
2010	• Washington DC City Council approves same sex marriage.

Note: * does not include legislative action on civil unions or court decisions addressing local actions on same sex marriage.
Source: National Conference of State Legislatures. Available at: http://www.ncsl.org.

constitutional amendment, popular with many socially conservative politicians, was also criticized as a transparent campaign strategy designed to energize the Republican Party voter base in the months leading up to the 2004 presidential election. The proposal failed to garner sufficient support in Congress that year. But this unsuccessful attempt at the federal level inspired 11 states to place their own defense of marriage amendments on the 2004 election ballot. After the dust cleared on election night, all 11 referenda had passed by healthy margins. By 2010, several other states had also adopted constitutional amendments banning same-sex marriage, bringing the total to 31 states (Figure 12.1).

As Webster, Chapman and Leib (2010) point out, a cursory examination of the spatial distribution of the voting results by state on such referenda point towards a regional geography that indicates higher concentrations of voter support for constitutional amendments in the American South than other regions in the US, particularly the 'deep south' states (Figure 12.2). Part of the reason for this

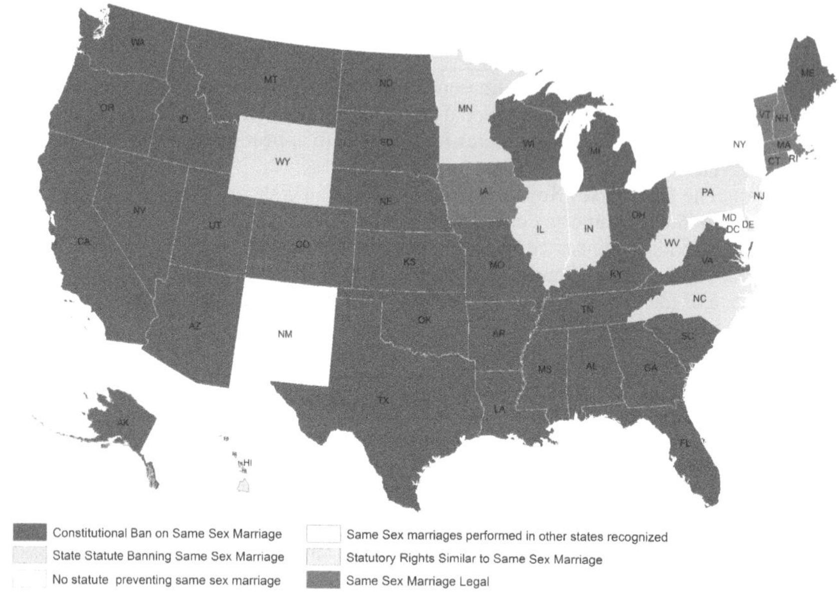

Figure 12.1 The Statutory Landscape of Same Sex Marriage in 2010

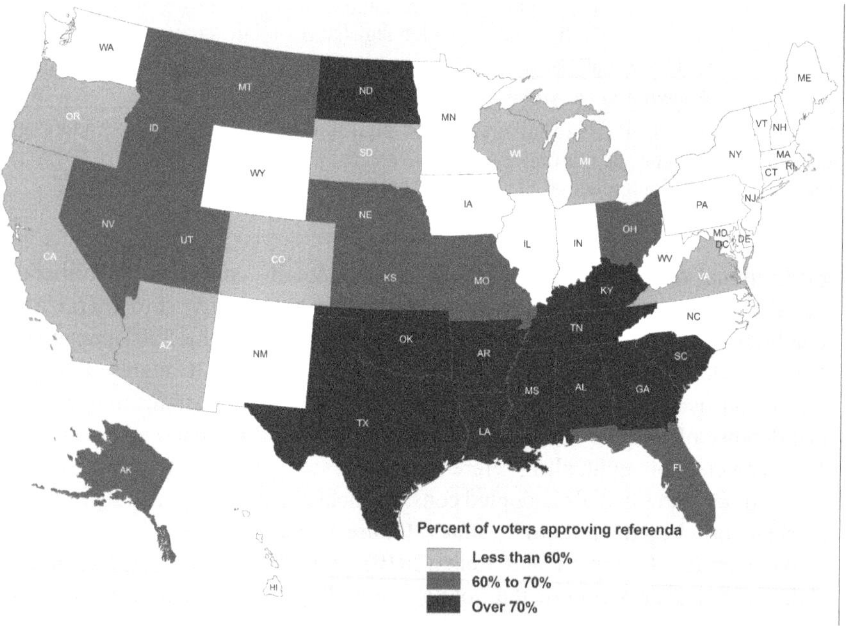

Figure 12.2 States with Voter Approved Constitutional Language Prohibiting Same Sex Marriage

is because southern states as a whole are disproportionately dominated by large numbers of fundamentalist evangelical protestants, helping to partly explain the lopsided support in favor of legislation banning same sex marriage. As a southern state, Florida's religious landscape is no exception. The state is also home to some of the most influential and well financed Christian Right political action lobbies in the nation, particularly in the central and northern parts of the state (Chapman, 2007). So it comes as no surprise that the debate leading up to the vote on *Proposition 2* was steeped with various religious jeremiads encouraging support for *Proposition 2*. But Florida's cultural and political geographies are also quite different from most other southern states in terms of its sheer population size, rates of in-migration from other regions of the country, and its complex ethnic and racial diversity. Indeed, the spatial outcome of the vote suggests multiple and complex political, cultural, and economic interpretations of the implications in instituting a constitutional ban on same-sex-marriage in Florida.

The Balkanization of the Marriage Rights Landscape in Florida

> Florida is like Italy in the 15th century. What you have here is a cluster of city-states, each with their own history and issues.
>
> Elizabeth Birch, former Director of the Human Rights Campaign, the nation's largest LGBT[4] rights organization

Long-time LGBT civil rights activists and those groups that actively oppose civil rights for sexual minorities are well aware of the countless culture war battles for sexual minority rights playing out in Florida for more than 30 years. These parochial struggles, initiated and debated at the local scale, also had major implications for the national LGBT rights movement (Stonewall Library and Archives, 2007). Of course any discussion of LGBT rights in Florida must begin with nationally known singer and orange juice spokeswoman Anita Bryant. Her (in) famous culture war crusade opposing the City of Miami's 1977 anti-discrimination ordinance, amended to protect sexual minorities from discrimination, was a very early extension of LGBT civil rights. Bryant's successful effort to repeal the ordinance though voter referendum less than a year later propelled Miami into the national spotlight, placing the city as ground zero for the many culture war battles on LGBT civil rights legislation that lie ahead, including the fight for marriage rights (Stonewall Library and Archives, 2007). As a direct result of the Miami debate, Florida's Governor Reuben Askew was inspired to take immediate action. Just one day after the ordinance's repeal he signed a bill into law prohibiting same-sex marriage in Florida (Boca Raton News, 1977). These early contributions to the culture war debates on same-sex marriage paved the way for other states to use Florida's legal, social, and cultural framework in opposing LGBT rights. Ten years

4 Lesbian, gay, bisexual, transgender.

later in 1997, the Florida Legislature took up the marriage rights issue once again by passing The Florida Defense of Marriage Act by an overwhelming margin. The bill's senate sponsor, John Grant, stated he thought it was 'great that [the Act] takes effect on June 4, right smack dab in the middle of Gay Pride Week' (quoted in Kanotz, 1998). The legislation was specific in its intent to strengthen the 1977 law, and was widely seen as a preemptive strike to fend off possible court challenges at the national level.

The vast majority of debates and legislative action granting civil rights for sexual minorities in the United States have taken place within local jurisdictions (Bailey, 1999). Collectively, these localized actions had major implications for public debate and lawmakers at the state and federal level. But a more accurate picture of support for or opposition to marriage rights for same-sex couples in Florida is revealed when place specific historical and geographical contingencies are considered. In other words, the same-sex marriage debate *within* Florida varied widely across space and time. On the surface it would appear that the citizenry of the sunshine state (and the political leaders that represent them) were mostly on the same page, firmly opposing same-sex marriage. But Florida's cultural and political geographies are much more complex and diverse than a first glance would indicate. Amidst the high profile public efforts by state and federal political leaders to legislate bans on same-sex marriage, LGBT Floridians began taking action on their own, and in their own communities.

As mentioned earlier, 2004 was the year when state and national debates on marriage rights hit a fever pitch. President Bush's efforts to federalize a ban on same-sex marriage were met with many local protests throughout the country by those opposed to legislative bans on same-sex marriage. The most high profile of all was not in Florida but in San Francisco, when Mayor Gavin Newsome defied California law by marrying hundreds of same-sex couples in the public corridors of city hall. Watching the spectacle from afar, same-sex couples in Florida began appearing at city and county courthouses all over the state, demanding marriage licenses. Ten same-sex couples filed suit in Ft. Lauderdale in March of 2004, seeking to overturn the 1997 Florida law and gain the right to legally marry. The lawsuit was the first formal legal challenge to state laws barring same sex marriage (Clary, 2004). That same week, seven same-sex couples showed up at Orlando City Hall in Orange County, also demanding marriage licenses (Santich, 2004). The deputy county clerk refused, handing the couples a copy of the state statute. The couples were not deterred, and were married later that day in a public religious ceremony sponsored by the First Unitarian Church of Orlando. A month later, hundreds of same-sex couples in Gainesville rallied at the Alachua County courthouse, also demanding the right to marry. And in a show of solidarity with LGBT Floridians, the city of Key West unanimously passed a resolution in support of same-sex marriage in Florida (Unitarian Universalist Association, 2004).

Florida4Marriage versus Florida Red and Blue

As these public spectacles in Florida continued to capture media attention, legislators in Tallahassee were again inspired to join in the fight by considering a state-wide voter referendum that would ban same-sex marriage by amending the state's constitution. Interestingly, the President's brother Jeb Bush, who was the governor of Florida at the time, was a bit more ambivalent, saying he would not rule out supporting such an amendment, but also added that it might be unnecessary since existing state laws accomplished the same purpose (Nevins and Davis, 2004). Without the active support of the governor, the proposal was never acted upon in the legislature. But behind the scenes, powerful factions of Florida's religious right were gearing up for a signature petition drive in order to get the initiative on the ballot in time for the 2006 election. A political action group calling themselves *Florida4Marriage*[5] was formed, lead by a cadre of activists from nationally known religious right organizations with regional headquarters in central Florida. Bankrolled by the state's Republican Party (Smith, 2006), their mission came up short when the group failed to collect enough petition signatures in time to get the referendum on the 2006 ballot (Florida Division of Elections, 2006). *Florida4Marriage* was not deterred, eventually succeeding in getting the required signatures to place the referendum on the 2008 ballot.

Lions, Tigers, and 'Activist' Judges from Massachusetts ... Oh My!

Local culture war issues often stem from how belief systems shape geographical imaginations, particularly those self-proclaimed 'gatekeepers' of community values that sense a threat to local prevailing values (Mitchell, 2000; Castells, 2004). These threats can come from inside the community or they can come from somewhere else. When the *Florida4Marriage* Chairman John Stemberger was asked by a Tampa Bay television reporter why he was pursuing a constitutional amendment to ban same-sex marriage when it's already illegal in Florida, he responded:

> Well, for the same reason Massachusetts needed one. The state had valid marriage laws, yet their judges just seemed to, with a stroke of the pen, erased this fundamental human institution that has served us since the beginning of time. So it's important, as 28 other states have done, that we enshrined the definition of marriage as a union of a man and a woman into our state constitution so a

5 *Florida4marriage* changed their name to *strongfloridamarriages* in 2009, focusing their attention on the divorce rate in the state by pushing for a "Marriage Preparation Act", which would require a state mandated waiting period and counseling for any couple wishing to get married in the state of Florida. For more information, see: http://www.strongmarriagesflorida.org.

> judge can't look to the state constitution to find some reason to overturn our valid marriage laws (WTVT Tampa Bay, 2007).

In this case it is 'activist judges in Massachusetts' who are most feared, playing the role of outsiders that pose a serious threat to marriage in Florida. Stemberger's othering of Massachusetts judges also depends on cloaking the issue with right vocabulary, in order to emphasize the gravity of the situation. Marriage, according to Stemberger, is a 'fundamental human institution' that has served us 'from the beginning of time'. Therefore it is necessary to 'enshrine' the very definition of marriage into a fundamental and timeless legal document that symbolizes the bedrock of American civilization (the constitution). If marriage ends up being defined by the tenets of civil law, or for that matter connected to love, then the seriousness of the 'threat' is minimized, and it becomes, according to Stemberger, easier to modify the rules for something that is just too sacred to change. Even more incendiary is the possibility that a judge, especially one that is not a Floridian, seems to be looking for a reason to use a 'stroke of a pen' as the weapon of choice to 'erase' all that is holy and good in Florida. Once this point is made (and it was made countless times during the lead up to the referendum vote), the implication is that marriage in Florida is exceptional and serious, while in Massachusetts it is a mockery and a sham. Indeed, the language written into Amendment 2 specifically prevents state judges from overturning the law to allow for same-sex marriage.

Supporters of Amendment 2 also saw an internal threat to marriage in Florida, one that existed within the borders of the sunshine state. Places such as Miami Beach, Broward County, and Gainesville had previously instituted domestic partnership registries as a legal alternative to prohibitions on legal marriage, viewing the definition of marriage as a matter of civil law that affords financial benefits, protects private property, and guarantees taken-for-granted privileges such as hospital visitation rights. The language that was written into Amendment 2 also addressed this 'threat', going further than most other state amendments in the sense that its passage would render *any* form of legal domestic partnership unlawful (including heterosexual partnerships). So the culture war battles over Florida's constitutional amendment redefining marriage would be fought on two fronts: fending off threats external to the borders of the state, as well those threats that lie within the state. And it was the voter referendum that would settle the matter in Florida once and for all.

Opposition to the measure was spearheaded by the political advocacy group FloridaRedandBlue (American Civil Liberties Union 2008). Through a well financed and savvy media campaign called "No on 2", FloridaRedandBlue's primary argument against passage of the marriage protection amendment focused on the restriction of civil and economic rights. For example, the amendment's passage would run counter to Article I of Florida's constitution (known as the

Declaration of Human Rights),[6] which sets a precedent for establishing civil rights rather than limiting them. In other words, it was argued that the amendment's passage would codify discrimination into the state's constitution by erecting a legal barrier that excludes an entire class of people from a state-sanctioned social institution, solely because of their sexual orientation. FloridaRedandBlue extended their argument beyond Florida's borders by stating that the amendment's passage would 'run counter to the proud US tradition of expanding, not denying, human rights via the Constitution' (*Miami Herald Tribune* 2008). This allowed the group to 'jump scale', taking advantage of the extra-scalar linkages inherent in local political debates by overcoming the constraints of their own geographically anchored struggles (Marston 2000).

Because the language of the amendment included a ban on legalizing other forms of domestic unions equivalent to marriage, the "No on 2" campaign argued that unmarried heterosexual couples would be in danger of losing their economic and legal benefits as domestic partners, such as hospital visitation privileges, funeral arrangement rights and health insurance policies that allow domestic partner benefits. The campaign emphasized that thousands of heterosexual unmarried senior citizen couples residing in the state take advantage of local domestic partnership benefits, and they would be especially hurts by the amendment's passage. In a series of widely debated television ads sponsored by *Fairness4AllFamilies*,[7] Floridians are introduced to senior citizens 'Ed and Clarice', long term domestic partners of over 30 years:

> Ed: We have a relationship based on love, trust, and mutual respect [Clarice leans over and kisses Ed]. In our domestic partnership it's very important that we are able to take care of each other.
>
> Clarice [her arm draped over Ed's shoulder]: As we get older we want to be able to take care of each other. We want to be able to visit each other in the hospital if it comes to that, and this amendment would take it away, those rights away.
>
> Ed [sitting close to Clarice on the sofa, holding her hand]: It's so important to vote against the amendment because it hurts real families, like our family. Please vote no on the so-called marriage protection amendment (Fairness4AllFamilies, 2008).

Opponents of the amendment also argued that domestic partnership registries in many Florida cities would also be threatened with expensive tax payer funded legal action, forcing them to shut their doors and cancel domestic partner benefits already in place. voicing her concern that the amendment's passage would be a

6 Available at: http://www.leg.state.fl.us/Statutes/index.cfm?Mode=Constitution&Submenu=3&Tab=statutes&CFID=176016536&CFTOKEN=57814895#A01.

7 Available at: http://www.myspace.com/fairness4allfamilies.

serious cultural setback for open-minded communities that currently have domestic partnership registries, University of Florida's women's studies professor Florence Babb stated that cities like Gainesville "would be losing a bit of its progressive orientation and its humanity (Dudash, 2008)."

Spatial Analysis of the Electoral Outcome

The overall spatial pattern of the vote on Amendment 2 broadly aligns with the socio-economic alignments found in most other work on the cultural politics of sexuality (see, for example, Brown, Knopp and Morrill, 2005; Chapman, Leib and Webster, 2007; Webster, Chapman, and Leib, 2010). The highest concentrations of the vote in favor of passage were in the panhandle counties, by far the most socially conservative and deeply religious portion of the state. Sixteen of the 17 counties where Amendment 2 passed by more than 80 percent of the vote were in this part of the state (Figure 12.3). The next highest concentration of votes in favor of Amendment 2 were generally found in the westernmost counties in the panhandle (collectively known as "The Redneck Riviera"), and the rural interior counties along Florida's peninsular 'spine'. With the exception of Gadsden County (which is majority African-American), these areas are overwhelmingly white and socially conservative, with large concentrations of evangelical Protestants.

Since approval of voter referenda in the state of Florida requires at least a 60 percent majority vote, the amendment was defeated in all of the southern coastal counties with large urban populations, particularly the South Florida megalopolis of Palm Beach, Broward, Miami-Dade, by far the most racially and ethnically diverse in the state.

The amendment also went down to defeat in the Tampa–St. Petersburg metro area, but only by a thin margin. The socio-demographics of these areas indicate a younger, more educated population than most other places in Florida. Additionally, these areas also have very large numbers of same-sex couples, particularly Ft. Lauderdale and Miami (US Census Bureau 2010). Surprisingly, voters in Sarasota County in Southwest Florida also voted down the amendment, a county that has been traditionally supportive of socially conservative Republicans and their causes.[8] Voters in Monroe County, which covers all of the Florida Keys, had the highest concentration of voters voting 'no', 52 to 48 percent. This is not surprising, since the City of Key West is one of the gay and lesbian "epicenters" in North America for both residents and tourists alike. Key West is also well known

8 Sarasota is the home of former Congresswoman Katherine Harris. Most recall Ms. Harris' controversial role as Secretary of State and Florida Republican co-chair in the immediate aftermath of the 2000 presidential election. Ms. Harris is nationally known for aligning herself with the most far right leaders and their causes, including the infamous Terry Shiavo controversy in 2005.

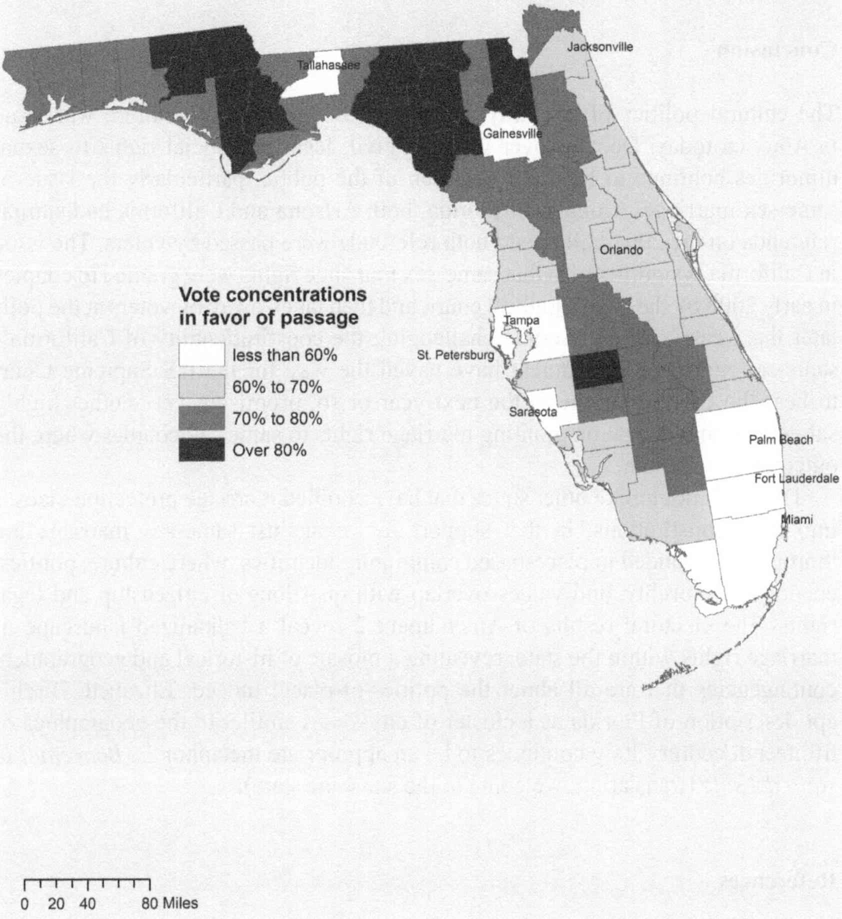

Figure 12.3 Distribution of Votes for Florida's Amendment 2 Banning Same Sex Marriage (2008)

for its long history of support for progressive political causes (Steinberg and Chapman 2009). Of particular note are the results in Leon and Alachua Counties, home to the state's two largest public universities, Florida State University and the

University of Florida, respectively. Though both counties are located within those regions within Florida that supported the amendment by healthy margins, voters in these counties voted down the amendment. In this regard, the voting results in counties are very similar to other large state university communities in other states that had similar constitutional amendments on the ballot banning same-sex marriage.

Conclusion

The cultural politics of sexuality is perhaps the number one culture war issue in America today. Debates over granting civil, legal and social rights to sexual minorities continue to hold the attention of the public, particularly the issue of same-sex marriage. Along with Florida, both Arizona and California had similar referenda on the 2008 ballot, and both referenda were passed by voters. The issue in California is significant in that same-sex marriage rights were granted to couples in early 2008 by the state's highest court, and then taken away by voters at the polls later that year. Various lawsuits challenging the constitutionality of California's same-sex marriage amendment have paved the way for the US Supreme Court to hear the case sometime in the next year or so, promising yet another highly salient national debate on granting marriage rights to same sex couples where the outcome is uncertain.

Florida is not unlike other states that have codified marriage protection clauses into state constitutions, in that support for or against same-sex marriage are thoroughly grounded in place-based community identities, where culture, politics, economics, morality, and values overlap with questions of citizenship and legal rights. The electoral results of Amendment 2 reveal a balkanized landscape of marriage rights within the state, revealing a mosaic of historical and geographical contingencies that are all about the politics-of-place. Indeed, Elizabeth Birch's apt description of Florida as a cluster of city-states similar to the geographies of fifthteenth century Italy continues to be an appropriate metaphor … *Benvenuti lo stato del sole* (translation: welcome to the sunshine state).

References

American Civil Liberties Union. 2008. "Florida Red and Blue launches 'SAY NO 2' campaign to fight Amendment 2, the Marriage Protection Amendment". Available at: http://www.aclufl.org/news_events/?action=viewRelease&emailAlertID=3388 [last accessed: October 14, 2010].

Bailey, R. 1999. *Gay Politics, Urban Politics: Identity and Economics in the Urban Setting*. New York: Columbia University Press.

Benjamin, J. and B. Nevins. 1998. Gay rights group seeks Florida's help: Tour building support for anti-hate law. *Florida Sun-Sentinel*, November 21, 1998. 1B.

Boca Raton News. 1977. Askew signs bill to ban gay marriage. June 9, 1977. 5A.

Brown, M., L. Knopp, and R. Morrill. 2005. The culture wars and urban electoral politics: Sexuality, race, and class in Tacoma, Washington. *Political Geography* 24: 267–91.

Castells, M. 2004. *The Power of Identity.* 2nd edition. Malden, MA and Oxford: Blackwell.

Christian Newswire. 2007. Available at: http://christiannewswire.com/news/811905071.html [last accessed: September 22, 2010].

Chapman, T. 2007. *Constructing the Moral Landscape through Anti-Discrimination Law: Discourse, Debate, and Dialogue of Sexual Citizenship in Three Florida Communities*. Ph.D. Dissertation, Florida State University. December 2007.

Chapman, T., J. Leib and G. Webster. 2007. Race, the creative class, and political geographies of same sex marriage in Georgia. *Southeastern Geographer* 47(1): 27–54.

Clary, M. 2004. 10 couples launch lawsuit over state ban on gay marriage. *South Florida Sun-Sentinel*. July 2, 2004. 1A.

Dudash, A. 2008. Opponents say Amendment 2 unfair, takes away benefits. *The Independent Florida Alligator*. University of Florida, October 31, 2008. Available at: http://www.alligator.org/news/local/article_2e87b84a-a50f-5661-a7c1-ebaeeba8769.html [last accessed: October 18, 2010].

Fairness4AllFamiles. 2008. Ed and Clarice Television Ad Campaign. Available at: http://www.youtube.com/watch?v=XMuHNIqU3G4 [last accessed: October 13, 2010].

Florida Division of Elections. 2006. Available at: http://election.dos.state.fl.us/.

Kanotz, M. 1998. For better or for worse: A critical analysis of Florida's Defense of Marriage Act. *Florida State University Law Review* 25: 2 (Winter).

Marston, S. 2000. The social construction of scale. *Progress in Human Geography* 24(2): 219–42.

Mitchell, D. 2000. *Cultural Geography: A Critical Introduction.* Oxford: Blackwell.

Scott, A. 2007. Amendment banning same-sex marriages closing in on ballot spot in November 2008. *Herald Tribune*. November 5, 2007. Available at: http://www.heraldtribune.com/article/20071105/NEWS/711050540# [last accessed April 10, 2010].

Miami Herald Tribune. 2008. Amendment 2: No. Editorial, October 11, 2008. A20.

Nevins, B. and J. Davis. 2004. Same-sex marriage ban gains in Florida. *Florida Sun-Sentinel*. November 10, 2004.

Rasmussen, C. 2006. We're no metrosexuals: Identity, place and sexuality in the struggle over gay marriage. *Social & Cultural Geography* 7: 807–25.

Santich, K. 2004. Gays say 'I do' after clerks say they can't: Seven same-sex couples denied marriage certificates tied the knot anyway. *Orlando Sentinel*, March 20, 2004.

Steinberg, P. and T. Chapman. 2009. Key West's Conch Republic: Building sovereignties of connection. *Political Geography* 28: 283–95.

Smith, A. 2006. GOP fuels gay marriage ban. *St. Petersburg Times*. January 12, 2006. Available at: http://www.sptimes.com/2006/01/12/State/GOP_fuels_gay_marriag.shtml [last accessed: August 2, 2010].

Stonewall Library and Archives. 2007. Days without sunshine: Anita Bryant's anti-gay crusade. Available at: http://www.stonewall-library.org [last accessed: April 4, 2010].

Unitarian Universalist Association. 2004. Available at: http://archive.uua.org/news/2004/freedomtomarry/florida.html [last accessed: October 15, 2010].

US Census Bureau. 2010. Available at: http://www.census.gov/ [last accessed: October 21, 2010].

Webster, G., T. Chapman, J. Leib. 2010. Sustaining the 'societal and scriptural fence': Cultural, social, and political topographies of same-sex marriage in Alabama. *The Professional Geographer* 62(2): 211–29.

WTVT Tampa Bay. Fox News 13 'Your Turn'. Aired December 12, 2007.

Index

For Product Safety Concerns and Information please contact our EU representative GPSR@taylorandfrancis.com Taylor & Francis Verlag GmbH, Kaufingerstraße 24, 80331 München, Germany